I0198790

Towards a Unified Cosmology

Reginald O. Kapp

Towards a Unified Cosmology

Copyright © 2019 Indo-European Publishing

All rights reserved

The present edition is a reproduction of previous publication of this classic work. Minor typographical errors may have been corrected without note, however, for an authentic reading experience the spelling, punctuation, and capitalization have been retained from the original text.

ISBN: 978-1-64439-198-3

CONTENTS

iii

iv

Author's Acknowledgment

Many friends have helped me with this book. They have directed my reading, corrected my misconceptions, primed me with facts, suggested turns of phrase. Often they have done so unawares, ignorant of the investigations I was pursuing. How can I acknowledge my debt to each of them? Who can keep account of all the written and spoken words from which his own thought has been distilled? So I can name in gratitude only those who have read all or parts of the manuscript and helped me to eliminate many defects literary, scientific and philosophical. These are my wife, Dr. Dorothy Kapp; Dr H. Gruneberg of the Department of Zoology and Comparitive Anatomy at University College, London; Dr Wilkinson of the Department of Biology at University College, Swansea; Mr.. W. B. Gallie of the Department of Philosophy at University College, Swansea; and Professor M. T. Smiley of the Department of Greek, University College, London.

PREFACE

The wider the range of a piece of research the less adequately can any one worker deal with each of its specialized aspects. Breadth and depth compete for his attention and cannot both secure the whole of it. I am only too well aware how particularly this truism applies to the study presented here. The very words 'Unified Cosmology' are both a challenge and a reminder that every conclusion arrived at has to be consistent with all the facts that have their appropriate place in every intellectual discipline. No conclusion here can be considered sufficiently tested by establishing its consistency with what is known in one branch of science only. Anyone who would make the most modest contribution to the unification of science should never claim to be writing specifically as a nuclear physicist, or as an astronomer, or as a relativist, or as a classical physicist, or as a mathematician, or as any other specialist whatever. When he does so he introduces a misplaced emphasis. He must never forget that knowledge acquired by himself in his own particular field of study is not necessarily more relevant, and may well be less so, than knowledge acquired by others in fields remote from his own. In order to find means of confirming or falsifying his conclusions, he must look for facts belonging to every branch of science to which he can obtain access. He must be prepared some day to be refuted by facts known to others, but of which he himself is still in total ignorance.

Appreciation of this solemn truth restrained me for many years from publishing more than a small fraction of the material that is presented in these pages. It was over thirty years ago that I first came to believe that the principle called here the Principle of Minimum Assumption deserved to be applied in the physical sciences with uncompromising consistency. I can trace this conclusion, or at least its clarification, to the impact made on me by Eddington's writings shortly after the First World War. Thus stimulated, I was led to notice how often in physics a scientist would, though perhaps hardly consciously, apply this principle and how fruitful the result invariably was.

Among many derivatives of the Principle of Minimum Assumption the one that soon claimed my particular attention was the one that I have since called the Hypothesis of the Symmetrical Impermanence of Matter; the hypothesis that matter is both originating continuously and continuously becoming extinct. If the Principle of Minimum Assumption was valid, I saw that all its consequences ought to be explored, including this one, and that exploration of

1

this hypothesis was likely to lead to new fields for research. But I did not regard myself as properly equipped for treading the path that was opening up before me. So I pointed it out to sundry scientists of my acquaintance in the hope that one or another of them would examine my suggestion critically. It seemed to me at the time that Symmetrical Impermanence would be difficult, though not impossible, either to confirm or to refute, but that the attempt to do so would lead to new insight, whatever the outcome. But none of those with whom I sought personal contact seemed to grasp the purport of my questions, and so I decided to approach a wider public by means of the printed page. I published the Hypothesis of Symmetrical Impermanence in 1940[1], though with a different name and in tentative terms. But still no one attempted to examine it.

Nevertheless, I felt encouraged when, some eight years later, Hoyle, Bondi and Gold published the hypothesis that they called Continuous Creation, for it agreed with my own view at least insofar as it postulated the random and causeless continuous origin of matter. Although these later authors could not accept my contention that a methodologically sound hypothesis about the duration of matter required that continuous origin be coupled with continuous extinction, they did help to prepare the scientific world for new insight into the relation between matter, space, time, causation.

This led me to return once again to the subject and also to renew my search for people willing to explore the implications of Symmetrical Impermanence. As a part of this quest I published a few of my own conclusions. They were necessarily tentative and were presented in the spirit of invitations to further research rather than as final, well-tested statements of fact. With these publications I approached a number of noted scientists and drew their attention to the promise of a solution of sundry problems that the subject seemed to offer to anyone who might feel inclined to pursue it, working either independently or in collaboration with myself. But again I failed to arouse any interest.

Time was passing and I was being forced reluctantly to realize that, unless I followed up my own suggestion, no one else was likely to do so. So, some six years ago, I began to give the subject more concentrated attention, regretting that I had neglected it for so long. But I never ceased to feel sure that a team would do the job better.

My method had to be adapted to the theme and yet to differ from that appropriate to research in a less extensive field. It is important to an understanding of what is presented here that this be appreciated. The aim of perhaps 90 per cent of all researches is to discover new detailed facts in a limited field. The starting point is experiment and observation. A collection of data is found that provides a useful addition to existing knowledge. Occasionally the research worker finds himself confronted with a puzzling

2

fact. This gives his work a new direction. An explanatory hypothesis has to be sought. It is an ad hoc hypothesis and, when found, has to be tested. If it stands the test, it becomes one of the generalizations of science; a new law, it is said, has been discovered. This pattern is so recurrent and familiar that any other is often deprecated. One must always start with a problem, it seems to be taken for granted, and devote oneself exclusively to the discovery of a law that will solve it.

Nevertheless, the familiar pattern cannot be applied here. Throughout the years while the theme of this book has been claiming some of my attention, it has not been my aim to find explanatory hypotheses. I have been seeking instead a literal and uncompromising justification in physics for the famous maxim: *hypotheses non fingo*. I have wanted, not to discover, but to test. And the assumption that I have sought to test is the Principle of Minimum Assumption itself.

My method for doing this has invariably been to consider first what inferences can be drawn from the Principle of Minimum Assumption and then to test these inferences for their compatibility with established knowledge. In other words, I constructed in thought the cosmological model that is implicit in the principle and then compared the model with actuality. Occasionally there seemed to be a contradiction between the inferred model and actuality, but further research always removed it. I was indeed surprised to find how many facts were explained that had hitherto had to be regarded as among those for which there is no explanation. The Principle of Minimum Assumption was found, again and again, to render ad hoc explanations unnecessary. This principle together with the Hypothesis of Symmetrical Impermanence, for which my justification had initially been purely methodological, came more and more to be also justified by their explanatory power.

Thus the Principle of Minimum Assumption has withstood every test to which I have been able to subject it. But if the principle is valid and universal in physics, it must govern every branch of the physical sciences. So the tests must continue. Whenever they are successful one may expect the experience recorded here to be repeated: something obscure will have been illuminated; new insight will have been gained. I should like this book to be regarded primarily as a more determined effort than my previous ones have been to encourage research workers to tread more often than they now do the path signposted with the word 'Testing' and not all to crowd along the path with the signpost bearing the legend 'To a discovery'.

Every theme calls for its own most appropriate treatment. When the theme belongs to a highly specialized field of study there is no excuse for a research worker who fails to comb the world literature on his theme before he commences his own research. In the thesis that he prepares at the end of it he is expected to quote from all the authorities whom he has consulted, to give a

3

detailed account of the existing state of knowledge on his subject, to assign priorities where they are due, critically to discuss all previous theories with mention of the names of their authors and the dates of their publication, to define precisely the point of view adopted by himself in relation to that of other workers in the same field.

It cannot be denied that the present study would be more reliable and better in every respect if I had done all this. But it is manifestly impossible for a single research worker to do so much when the subject is as comprehensive as it is here. I had to face the dilemma of either presenting something that I knew to be imperfect or of presenting nothing at all. So I decided that my aim should not be perfection but that I could, nevertheless, stimulate thought along new lines.

If the author of new theories waits, moreover, until he has found a satisfactory answer to every question that arises from the theories he will wait for ever. Would that this were more widely appreciated. It is a common observation that those who dislike a new theory but are unable to refute it seek only too often to have it ignored by drawing attention to some problem that remains unsolved. 'No answer has yet been provided to such-and-such a question', they say. 'And so the theory must be wrong. It is best forgotten.'

I cannot hope to escape this shallow kind of comment, but I may forestall if it I define the status of my theories as it appears to me. I show in this book that Symmetrical Impermanence can be justified on methodological grounds and provides answers to an impressive list of questions in a variety of scientific disciplines. This does not suffice to prove it true but it suffices to warrant a serious attempt to answer those questions that arise out of my theory.

The attempt may lead to falsification of the theory. It has happened often in the past. But a theory cannot be falsified by showing that it is incomplete. It is high time that this was pointed out by someone who has not yet suffered from the kind of criticism that accuses him in effect of not knowing all the answers. For what constitutes indignation at the way other men's theories have too often been received in the past may appear as touchiness about the reception of one's own. This is why I want to point out that ignoring is no substitute for refuting; and I want to point it out before, rather than after, it has happened to me.

Some of the problems of presentation that arise from the great breadth of the subject are worth mentioning.

There was the question how to quote chapter and verse for those statements for which I could claim the backing of authority. The usual method is by citations, footnotes, and a comprehensive bibliography. But for the present purpose I decided to dispense with the ostentatious support of authority. So long as I avoided any suggestion that the discoveries of others were my own, I could not mislead by omission of an author's name and I might do an injustice if, through ignorance, I assigned a wrong priority to anyone. I have, therefore, been very sparing of references and footnotes. If anyone says that these omissions prove my ignorance, he will be right. If he says that they prove my ignorance in some particular instance, he may be wrong.

But ignorance has not always been my reason for omitting the names of authorities whose theories I have to refer to. The number of rival theories seems to be greater in cosmology than in most sciences with the exception of psychology. In cosmology there also seems to be a tendency to do more publishing and less critical examining of theories than in most branches of science. This explains why in the past there have been so many theories about the sun and the planets, the stars, the nebulae, space and the universe in general. Each has at one time had the backing of high authority; it has been widely accepted for a few years or even a few decades; it has then been rejected because of some rather obvious defect; another theory has taken its place. Rejection has not always been based on a more recent discovery; often the defect could have been noticed at the beginning.

What held for the past holds, unfortunately, for the present day. In the course of my reading I have met many theories in the field of cosmology that are quite recent and yet, I am afraid, quite untenable. But they are so close to my subject that I cannot ignore them. Without extensive research into the literature of the subject I cannot always know for certain who their original authors have been, nor whether they still sponsor them, nor how strong their present backing is. So to quote from the particular passages that I happened to have found might do someone an injustice. To discuss such theories while coupling them with the names of particular authorities would introduce an undesirable polemical note. It would also be unkind. I have compromised by introducing such theories with some such form of words as 'one might think for a moment that', or 'the possibility has to be considered that'. Those who have not met the particular theory before may be misled into thinking that I have set up a dummy in order to knock it down; but that is here the lesser evil.

Technical terms caused me some anxious thought. In an investigation that is limited to a particular branch of science they should always be used. But those adopted in one branch are unfamiliar to specialists in another, which has often precluded their use here. For this reason I have, for instance, spoken cumbersomely of 'the formation of mountain ranges and depressions', instead of using the neater and more precise geologist's word 'orology', and in discussion of relativity I have found means of avoiding the term 'invariant'. Sometimes I have deliberately said something in a way that I know an expert in the branch in question would not adopt. If, thereby, I risk irritating the expert, I am not likely to mislead him and I have tried to find the method of presentation that would give a maximum of information to the non-specialist.

Those parts of the book in which relativity theory has to be discussed have raised particularly difficult problems of presentation. This is because only a small fraction of those whom I am addressing here are likely to be familiar with the mathematics of general relativity, though some of them may have no difficulty with special relativity. The mathematics is not as difficult as is often supposed, but the notation is unfamiliar and does not occur in most other branches of science. Many scientists can do all their work without ever using this notation. The letter symbols used by relativists constitute a highly condensed kind of shorthand in which one letter represents a great deal. It need not surprise that those who do not use the notation daily easily lose sight of the physical concepts for which the letter symbols stand. Even expert relativists do so sometimes, which explains why there is still some doubt about the correct physical interpretation of certain aspects of general relativity. It also explains the occasional reluctance of relativists to give any physical interpretations at all. But here I have repeatedly found myself obliged to do just this. Interpretations given by others have all too often failed to secure general acceptance and I cannot hope for a better fate for mine. From correspondence and conversation I have been led to expect, in particular, somewhat violent opposition to my claim that, in the relativity equations, the letter symbol that represents mass must usually be interpreted as representing inert and gravitational mass only and that great care must be taken to ensure that the same letter symbol is not wrongly interpreted as also representing attracting mass. I have written with the express purpose of inviting critical comment and have to add that, while the majority of those whom I am addressing are competent to provide critical comment on most of what is presented here, only a minority of experts in general relativity are competent to do so about my interpretations of the relativity equations. It places rather a big responsibility on these few.

I wish to acknowledge valuable help with the presentation of my material that has been given me by Mr.. B. C. Brookes and Mr.. C. R. Howe. I also wish to acknowledge much scientific advice and information from Professor Schieldrop of the University of Oslo, to whom I showed an early draft.

Chapters I and 2 and passages from other parts of the book in which space is discussed have been published in *The British Journal for the Philosophy of Science*. I am indebted to the editor and publishers of that journal for permission to publish them here.

REGINALD O. KAPP

SUMMARY

Part One - **THE UNIFICATION OF PHYSICAL SCIENCE**

The hypothesis is defended that the laws of physics are not restrictive in the sense in which laws in statute books are. The laws of physics, it is claimed, neither require nor prohibit any specific number, event, condition, property, configuration or other feature. They permit everything to occur that is logically consistent with all observable facts and that is, in this sense, logically possible. This can be expressed by saying that every valid generalization in physics can be so stated that the term 'any' or 'either' occurs in its formulation: Examples are - 'A solar system may contain any number of planets. A nuclear particle may carry any number of unit charges up to the stability limit. Space may conform to any geometry. An indivisible particle may carry either a positive or a negative unit charge.' It is shown that physicists have often acted on this hypothesis in the past and that, when they have done so, progress has been made towards the unification of physics. Conclusions have, nevertheless, sometimes been based on the alternative hypothesis that the laws of physics are of the statute book kind, though perhaps inadvertently. These conclusions have had to be abandoned later in favour of non-specific ones. In this respect, it is claimed, physics differs basically from all other disciplines, including biology. The hypothesis defended here is given the name 'Principle of Minimum Assumption'. It is declared to be the most basic of all the principles of physics. If it is applied with uncompromising consistency, it precludes every *ad hoc* explanatory hypothesis of a specific kind. For if the Principle of Minimum Assumption holds for the whole of the physicist's universe of discourse, it must be theoretically possible to infer all other generalizations in that universe from this principle without the need for any other hypotheses.

Part Two - THE PAST AND FUTURE DURATION OF MATTER

It is shown that there is a choice between nine different assumptions about the duration of matter and energy. They are all hypothetical and five of them have had the support of various authorities. These hypotheses are discussed critically and tested for their conformity to the Principle of Minimum Assumption as also for their consistency with known facts. It is shown that, with one exception, each of them can be supported only with the help of a substantial number of additional ad hoc hypotheses. The exception is the hypothesis that any elementary component of the material universe may originate at any moment of time and become extinct at any moment of time. It is called the Hypothesis of the Symmetrical Impermanence of Matter. It is shown not to be an independent hypothesis but an inference from the more basic one called the Principle of Minirnun Assumption.

The Hypothesis of the Symmetrical Impermanence of Matter is tested for its consistency both with what is known about causation in physics and with the conservation laws. It is shown that origins and extinctions have to be regarded as uncaused events in the same sense in which the disintegration of a radio-active atom is regarded as an uncaused event. To accept Symmetrical Impermanence is therefore not to change the physicist's present conception of the nature of causation. It is also shown that this hypothesis is not inconsistent with the conservation laws when these are given the form that expresses precisely the use to which they are put; but these laws have come often to be imprecisely worded and can then seen to refute Symmetrical Impermanence. It has to be emphasised that, according to Symmetrical Impermanence, origins and extinctions are absolute and not accompanied by the conversion of energy from one form into another. Extinctions are, therefore, not accompanied by release of any energy.

Part Three - THE ORIGIN AND EVOLUTION OF GALAXIES

According to the Hypothesis of the Symmetrical Impermanence of Matter, new matter is originating everywhere. In a region large enough to be a fair sample of the whole material universe the rate of origin per unit volume is constant. Hence most origins occur in extragalactic space. The resulting atoms of hydrogen find themselves in the very faint gravitational potential gradients that are caused by existing nebulae. The atoms experience accelerations that, though very small by terrestrial standards, continue for sufficient periods of time to lead to large velocities and to displacements of matter over large distances in large quantities. If Symmetrical Impermanence is true, these faint gravitational potential gradients determine the distribution of matter in extragalactic space. The gradients are discussed in detail and the expression used to describe them is 'astronomical landscape'.

It is shown that the detailed features of the astronomical landscape are such that new clouds must form in extragalactic space at finite intervals of time and eventually acquire the characteristic shape of the spiral nebulae, including the spiral arms. One should expect the formation of these to be accompanied by turbulence in a very large quantity of very tenuous gas and to be revealed by a radio-telescope. Some of the radio-stars that have been observed in regions remote from any visible nebula may be incipient nebulae at the stage at which their spiral arms are formed.

Part Four - GRAVITATION

It is shown that our understanding of gravitation is more defective than that

8

of most natural phenomena and eight questions of major significance are listed to which answers cannot yet be provided. Among these is why an accumulation of inert mass is the source of a gravitational field. A new theory of gravitation is provided that gives an answer to this questtion as well as to the other seven. This theory is based on two physical principles. One is the Symmetrical Impermanence of Matter, the other the relativistic concept of curved space.

If the new theory is true, a particle does not carry an extensive gravitational potential gradient around with it, as has hitherto been supposed. The gradient occurs only as a consequence of the extinction of the particle and as a momentary pulse. Gravitation is quantized and could be described as the swan song of matter and not, as supposed by tradition, as its signature tune.

It is pointed out that the formation of stars from a tenuous gas and their rotation are both more difficult to explain than is often supposed. Indeed, no tenable explanation has hitherto been provided for either. It is shown that, if gravitation is quantized and the pulses in a very tenuous gas are significantly intermittent, the formation of stars as well as their rotation and the rotation of the spiral nebulae can be accounted for.

Appendices

It is inferred from sundry known facts that the half-life of matter is of the order of 4×10^8 years. This means that the mass of the Earth, and with it the value of g, is and has been continuously becoming smaller.

Astronomical, geological and biological implications of this finite half-life of matter are worked out and it is shown that it helps to explain a number of facts that have, hitherto, defied explanation. If the mass of the atomic nucleus is a region of intensely curved space the hypothesis of a nuclear force in order to explain the cohesion of the nucleus is shown to be unnecessary. It is concluded from this that the units becoming extinct are electrons and complete atomic nuclei.

PART ONE

THE UNIFICATION OF PHYSICAL SCIENCE

Chapter 1

The Concept of Unification in Physics

1.1: *The Meaning of Unification*
This study is the search for a unified cosmology. Before it is begun clarity must be reached as to the meaning attached to the word 'unification'.

As understood here it constitutes the replacement of many specific laws, principles and hypotheses by a smaller number of more general ones. An example of the process, which is frequently quoted and which stands out as pre-eminent, is the unification achieved by Newton.

Before his day there were no general laws of mechanics; there was only a variety of specific laws, each applicable to a specific mechanical system. It was believed that a specific law, applicable only to planets, required these to move in elliptical orbits; that a quite distinct specific law, applicable only to pendulums, required their period of swing to bear a specific relation to the length of the pendulum; that yet another specific law, applicable only to vacuous spaces, required these to be filled. If there were such a thing as a Cosmic Statute Book, this would have had to contain, according to the pre-Newtonian view, separate entries under the respective headings Planets, Pendulums, Vacuum. The book would have been a bulky one.

But Newton showed that many such specific laws were implicit in other more general ones. A large number of observed facts could be inferred from his laws of motion and gravitation. If the Cosmic Statute Book contained these all-embracing laws, there would be no need for further entries to say that planets shall move in elliptical orbits, that the period of swing of a simple pendulum shall be proportional to the square root of its length, that a projectile shall have a parabolic path, that a vacuous space shall be filled: Newton, it might be said, did much to whittle down the Cosmic Statute Book.

Another way of describing the same achievement is to say that Newton 'explained' a large number of facts. For in physics a fact is explained by showing how it can be inferred from something more general. All explanations there are steps in the direction of greater unification.

Since Newton's day the unifying process has extended into more and more

11

branches of physical science. The first and second laws of thermo- dynamics have had great unifying power. From them much can now be inferred that would otherwise have to be attributed to specific, ad hoc, laws. The relation between the once quite distinct subjects of electricity and magnetism is found to be so close that they are now considered as one subject. A study of the relation between chemical reactions and thermodynamics, as also of that between chemical reactions and atomic structure, has led to the new branch of science called physical chemistry. This has taught us that chemical processes and properties are implicit in atomic structure. At one time it must have appeared that an entry in the Cosmic Statute Book would be necessary to say that hydrogen shall combine with oxygen and form a substance with the properties of water. But we now know that such a clause would be redundant. Physical chemists can tell us that, provided there be atomic nuclei with, respectively, one and eight positive unit charges, the rest is assured. One can infer with the help of certain general laws that atoms having such nuclei must combine and that the resultant compound must have the properties of water. This unification is making it possible to explain more and more chemical facts in terms of atomic structure.

1.2: *Unification leads to Predictions*
A most valuable feature of unification is that it enables one to replace observation and experiment by inference and calculation. Galileo could discover the law of the pendulum only by observing pendulums; but after Newton anyone who had never seen a pendulum could have discovered the law. He would have done so by the reasoning that is now taught in schools, where the formula is deduced from first principles. Similarly the elliptical orbits of planets could have been discovered before Newton's day only by making careful observations of successive positions of planets. But now a person who had lived all his life under a blanket of impenetrable cloud and learnt for the first time that there was a massive sun surrounded by less massive bodies could predict that, when the cloud lifted, one would observe the less massive bodies to move in elliptical orbits. There was a time, again, when the chemical properties of substances seemed to be discoverable only by observing those substances. But this is not necessary today. Long before the element hafnium had been found chemists did not only say that there was such an element; they also predicted its properties. The properties are implicit in the general laws that chemists have discovered and can be inferred from these laws.

It is this kind of unification that has made the rapid progress of technology possible. If every chemical substance had to have a clause in the Cosmic Statute Book definining its properties, chemists would have to make the substance and submit it to a laborious series of tests before they knew what the properties were. But the properties are implicit in general laws. If these are known, the properties follow automatically. Hence it is a commonplace of chemical research to predict the properties of a new compound before making it.

It is the same in all other branches of technology. Without a unified physics one would have to make a gun and fire it before one could know what path the projectile would take. One would have to make and test a bridge in order to discover its strength. One would have to make and test every new kind of engine before one could determine its thermal efficiency or the critical speed of its shaft. But the technologist's aim is to substitute inference for observation. Doing this, he can predict the performance of guns, the strength of bridges, the efficiency of steam engines while they are still in the blue-print stage. In technology, tests, observations, experiments, do not only serve the purpose of facilitating predictions but, often instead, of verifying them and of correcting errors and oversights. This can be done because what is predicted is implicit in general and known laws and principles.

1.3: *The Search for Greater Unification Continues*
By the process of unification the whole of physical science is gradually being fashioned into one complete and consistent structure of thought in which the various parts bear a logical relation to one another. Mechanics, electricity, magnetism, thermo-dynamics, chemistry, heat, light, sound, to mention only some branches, have been brought under one common roof.

During the present century we have seen two conservation principles, those respectively of energy and mass, united. Einstein has established in the general theory of relativity a connection between gravitation and space and has thereby brought space under the common roof with the rest of physics. There has, further, been the formulation of the very basic and comprehensive law according to which physical changes cannot be by indefinitely small amounts. This law forms the foundation of the quantum theory and has brought under the common roof a large number of observa- tions that previously seemed to be isolated and each to require its own clause in the Cosmic Statute Book. Predictions of all sorts are being based every day on the principle that all physical changes are quantized.

The search for greater and ever greater unification continues, but with varying success. One of the failures is worth mentioning because it illustrates the nature of the problem. Three different types of field of force have been observed: magnetic, electrostatic, and gravitational. Something is known about how the first two are related to each other, and one commonly speaks of them jointly as the electromagnetic field. But yet they remain distinct from each other and quite distinct from the gravitational field, which has been shown by Einstein to be a region where the geometry of space-time differs in a specific way from the geometry of Euclidean space. The difference can be expressed in a mathematical formula.

The hypothesis is near at hand that the magnetic field is also a region where the geometry of space-time differs from Euclidean geometry, though in a different way, and that the electrostatic field represents a third departure. If so, one might expect to be able to generalize Einstein's relativity equations in such a way that they would represent any kind of field. If that could be done, one specific value of a term in the equation would define the gravitational field only, another the magnetic field only and a third the electrostatic field only. Each field would then appear as a special case of something common to them all; its properties could be predicted from the great sweeping law that was applicable to all fields; magnetic, electrostatic and gravitational fields would be brought under a common roof. The attempt to achieve this has been called the search for a unified field theory.

Assiduously though it has been conducted by a number of scientists, of whom Einstein was one, the search has so far led only to disappointment, It is impossible to say yet whether the failure is due to the inherent difficulty of the subject or because the search has taken a false hypothesis as its starting point, i.e. that all fields of force have enough in common for them to be represented in terms of the geometry of space-time. Yet, in spite of the apparent reasonableness of this assumption, it may not be true. Electrostatic and magnetic fields may be so different from gravitational ones in their nature, their effect, their cause, that they cannot be represented in any comparable terms at all. Some other hypothesis, one that has not yet been formulated or even thought of, might prove a better starting point for bringing electromagnetism and gravitation under a common roof.

Be that as it may, no attempt will be made here to succeed where Einstein and others have failed. The present study is in no way concerned with the search for a unified field theory, desirable though it is that the search should continue. But the example will help to define the scheme according to which unification in physics is achieved.

1.4: *How Unification is Achieved*
In the process of unification in physics a general principle is first found. Examples are Newton's laws of motion, the great conservation laws, the principle of least action, the principle according to which all observable physical changes are quantized, the principle of the equivalence of mass and energy, the principle by which the chemical properties of substances are related to the number of electrons that surround the nuclei of the constituent elements.

At the next step towards unification various phenomena are shown to be implicit in one or other of these principles. They can therefore be inferred from them and so could be struck off the Cosmic Statute Book as redundant.

Sometimes the phenomena are observed first and the principles are found later. The principles are then said to explain the phenomena. Thus the observed behaviour of planets was explained by the laws of motion and gravitation. Similarly, attempts to make a perpetual motion machine failed for unexplained reasons until the principle of conservation of energy provided the explanation.

At other times the principle is found first and some phenomenon that is implicit in it is described before it has ever been observed. In such instances it may, or may not, be observed later. One then says that the phenomenon is predicted by the principle. Engineers, as mentioned already, follow this course as a matter of routine. They invent and design new kinds of machines on the basis of the great sweeping principles of physics and they predict their performance. Observation and experiment come laterand not to test the principles but to test the soundness of the designer's reasoning. In physics, too, it sometimes happens that a phenomenon is predicted as an inference from a general principle before it has been observed. The properties of hafnium have already been quoted as an example. The discovery of Neptune by Adams and Leverrier and of Pluto by Lowell are other examples. But physicists work most often with things that they are observing at the time. Their concern, unlike that of the machine designer and the industrial chemist, is more often to explain observed effects than to predict those that will only later become observable.

The striving to bring ever more phenomena under the common roof, to unify the whole of physics, is, of course, not the whole of the physicist's work. Indeed most research workers are concerned only with the discovery of the detailed facts, qualitative and quantitative, of the physical world; and necessarily so, for we still have much to learn about the laws of mechanics, heat, light, sound, electricity and magnetism; about the physical and chemical properties of solids, liquids and gases; about the macrostructure and the microstructure of the material universe, about the positions and movements of the heavenly bodies. But, nevertheless, it is worth stressing that much of the thinking done by most physicists is directed towards the discovery of generalizations and that it is on these as much as on collections of observed facts that physical science is based.

The distinction between the search for isolated facts and the search for unifying generalizations is well illustrated by the elliptical orbits of planets. We can now understand why these orbits could not be explained before Newton's day. It was because of a wrong outlook. During and for some time after the Middle Ages the notion was prevalent that every phenomenon was the result of what might be called a distinct act of legislation, that it was ensured by what I have metaphorically called a separate clause in a Cosmic Statute Book. Those who held this view were bound to think that it was idle to ask why the planets moved in the observed orbits. The acceptable answer was that such orbits were a legislative requirement, about which no further questions could or should be asked.

But we can now realize that, if the planetary orbits could not be explained before Newton, it was not that they were inexplicable. Nor was it that not enough was known about planets. It was that not enough was known about mechanics. No further astronomical research, no careful observation of the orbits, no precise measurement could have provided the explanation. But Newton's laws of motion and gravitation did so. In other words, scientists found the answers to some specific questions about planets only when they had found statements that were general enough to apply to all ponderable objects. In medical metaphor, ignorance about planets proved to be, not the disease itself, but a symptom of the disease. Newton followed the course of a doctor who seeks to treat the disease rather than the symptom.

Similarly our inability until a short while past to explain why given chemical substances react in the observed ways proved to be, not due to our ignorance of the substances, but to our ignorance of the more general subject of atomic structure. The great generalizations on which modern chemistry is based could not have been discovered by work conducted only in the field of chemistry.

These considerations are relevant to the present study because I propose to demonstrate here the great explanatory power of the generalization that is reached when one pursues the search for a unifying principle in physics with uncompromising persistence. It will be shown in the next chapter that one then reaches a very comprehensive principle, one that is applied by physicists on occasion but which has not been given the status it deserves. It will be called the Principle of Minimum Assumption and will be described fully in the next chapter. It will be shown in the remainder of this book that one can infer from this principle, and without the need for any further hypotheses, a number of cosmic phenomena that have hitherto eluded explanation. The selected examples will be the expansion of space, the occurrence and detailed structure of nebulae, and the familiar observation that every large accumulation of inertial mass is the source of a gravitational field.

Chapter 2

The Principle of Minimum Assumption

2.1: *Occam's Razor*

Any search for greater unification of physics must have a reasoned beginning; it must be conducted methodically; the unifying principle sought must conform to the requirements of scientific method. So let the starting point of this inquiry be one of the commonplaces of scientific method the rule of economy of hypotheses, sometimes called 'Occam's razor'. This says that when more than one explanation of an observation is available one must provisionally choose the one that involves the least number of assumptions. The rule is so well known and so generally accepted that there is no need to illustrate it by examples. Most of us never doubt that it is a good rule, although we may differ as to how rigidly it should be applied. To assess its value we must first consider its nature and then the place that it occupies in scientific research. We shall find that in the history of physics the importance of this rule is very great indeed.

The rule of economy of hypotheses is one of the canons of scientific method. It tells the scientist what to do when more than one explanatory hypothesis is available, but it does not guarantee that the recommended choice will be justified by events. This is evident from use of the word 'provisionally'. By advising that the minimum assumption be made provisionally the rule allows for the possibility that another hypothesis, one that does not meet the criterion of minimum assumption, may have to replace it some day.

The great principles of physics are in a different category. They are not mere rules of procedure but statements about the very nature of the physical world. They are so well established that the word 'provisional' is omitted from their formulation. The principle of conservation of energy is an example. It is concerned with the energy in a given system, which may be energy of motion, of position, of chemical structure, of mass. The principle asserts that, provided the system be self-contained, the total quantity of energy in it is constant. Many conclusions can be inferred from this great principle. One of them is that a perpetual motion machine is impossible. So the principle of conservation of energy suffices to refute a person who claims to have invented such a machine. It would be highly exaggerated caution to tell him that his idea has to be rejected provisionally; one rejects it outright. One does not say that it is unlikely that the machine will work; one says that the principle of conservation of energy proves the inventor's idea to be wrong.

It is not so with a statement that violates the rule of economy of hypotheses. Almost every week letters appear in the daily press and articles in scientific journals, papers are presented to learned societies, in which hypotheses are put forward that involve more, sometimes much more, than a minimum assumption. One may deplore such hypotheses, but it is not customary to use Occam's razor to prove them wrong.

Here is a significant difference between a principle and a rule of procedure. One can *refute* a statement that violates the principle of con- servation of energy, but one can do no more than *deprecate* a statement that violates the rule of economy of hypotheses.

In making this distinction do we give a sufficient status to the rule of economy of hypotheses ? It depends, I am venturing to suggest, on the discipline with which one is concerned. In history, in biology, in the social sciences, the rule can be no more than a useful guide to procedure; state- ments that conform to it can be accepted only provisionally; they may eventually have to be replaced by statements that violate the rule. But I wish to make the bold claim here that, in physics, the rule of economy of hypotheses can be so expressed and defined that it acquires a status far higher than the one usually accorded to it; I wish to raise it from a mere rule of procedure to one of the great universal principles to which the whole of the physical world conforms. At this level it would be worded as follows: *In physics the minimum assumption always constitutes the true generalization.* It needs a name so I propose to call it the *Principle of Minimum Assumption.*

This claim is, itself, an hypothesis and has to be justified. I propose to do so in subsequent chapters by showing that a unified cosmology is achieved by the consistent, uncompromising and methodical application of the Principle of Minimum Assumption to theories about the past and future duration of matter. But before I do this I shall have to discuss the nature and meaning of the principle and show that it is applied by scientists already and more often than is always appreciated.

2.2: *What is a Minimum Assumption?*

One sometimes hears the remark 'that is a big assumption', just as one hears 'that is a big lie'. The implication of such colloquial habits of speech is that quantitative distinctions can be made between different assumptions and between different lies; that one could arrange a collection of assumptions or lies in a row, with the biggest at one end and the smallest at the other; that the magnitude of assumptions and lies could be expressed in units, like those of temperature, hardness and other measurable quantities.

It may be so for all I know. But I am not concerned here with the grading of assumptions according to size. I am concerned instead with the search for a criterion by which an assumption that is defined as a minimum one can be clearly distinguished from assumptions that cannot be so defined. I do not think that the criterion is hard to find. I think it is whether the assumption is specific or not; so I shall define a minimum assumption as one that is completely unspecific. What this means can best be explained with the help of some examples. The first of them will be deliberately chosen to be extreme to the point of absurdity.

A young man who has had a predominantly humanistic education has been fascinated by a popular book on astronomy and has become enthusiastic about what he calls the beauties of science. He has read in his book that most stars do not have planets but that a very small proportion of them do and that these form solar systems like our own. The total number of stars is great, he learns, and so even the tiny fraction that have solar systems amounts to millions of stars.

He has also read in earlier books that planets are sublime bodies, constrained by their noble natures to move in orbits of geometrical perfection. In doing so, he has gathered, they produce a lovely harmony, known as the music of the spheres. He reaches the not unnatural opinion that, where perfection is displayed in terms of geometry and music, it must also be displayed in terms of number, so he arrives at the conclusion that every one of those millions of distant solar systems must have a pleasing number of planets. He can well believe that this may be the mystical number seven, or the virile number nine, or occasionally perhaps the round number ten. But he feels sure that no solar system can be cursed with the unlucky number thirteen.

An astronomer friend reproves him for his unscientific outlook. To believe that the number thirteen is precluded is, he says, a big assumption and an unjustified one. He tells the young man that some solar systems do have thirteen planets. The young man remains puzzled. Why, he asks, is it a big and forbidden assumption to believe that no solar system has thirteen planets and a small and permitted assumption to believe that some solar systems do have thirteen planets?

The question is not a silly one. It must not be dismissed with a shrug and a smile. It is basic to scientific method and it behoves us to find a clear and direct answer to it.

The answer is not that solar systems with thirteen planets have been observed. The astronomer admits that his belief in the existence of such solar systems is an hypothesis. The resolving power of our best telescopes is not sufficient to reveal any solar system but our own. So far as observation goes we have no proof that there is any other solar system at all.

19

Nor is the answer that there is a law of physics by which some solar systems are required to have thirteen planets. The reason for the astronomer's belief is, on the contrary, the very *absence* of any known law to require that solar systems shall have a specific number of planets. Here the minimum assumption is the unspecific one, i.e. that *any* number of planets can occur. This is therefore the assumption that, in conformity with the demands of scientific method, is made by our astronomer. He would not feel justified in making any other.

From the above little story one may learn the operative word by which to recognize a minimum assumption. It is 'any'. It need not apply only to numbers, but can also apply to quantities, properties, relationships, configurations, to any feature that one likes to mention. When there are only two alternative possibilities the grammatical substitute for 'any' is 'either' as, for instance, when there is a choice between the positive and the negative sign. So a minimum assumption can be recognized by the use in its formulation of the words 'any' or 'either'.

In practice it is not difficult to distinguish between a minimum assumption, as just defined, and one that is not a minimum one. But the question remains whether, when one has recognized a minimum assumption, one is always justified in making it. In the example of the number of planets in a solar system one cannot be sure whether the minimum assumption is the true generalization or not. For it is impossible to prove by observation that solar systems may have *any* number of planets. But there are many occasions in the history of physics when the minimum assumption has proved to be the true generalization; and I can recall no occasion in the history of physics when it has not been so. Let this be demonstrated with the help of some real examples.

2.3: *Use of the Principle of Minimum Assumption in Physics*
Minimum assumptions have not received as much attention as one might have expected in the philosophy of science. The distinction between specific and unspecific assumptions is not a textbook subject; and so but little has hitherto been done to clarify thought about it. But the history of science shows that specific assumptions have, in the past, been made time and time again; that they have been treated as generalizations about the nature of the physical world, and have eventually had to be replaced by unspecific ones. Whenever this has happened new light has been shed on a wide range of subjects; the unifying and explanatory power of the unspecific assumption has been demonstrated.

Thus it was assumed at one time that a specific law, applicable only to planets, constrained these to move in elliptical orbits. But Newton replaced the hypothesis that the planetary orbits were the consequence of a specific law by the hypothesis that they were the consequence of the circumstances in which the planets found themselves. The assumption that certain bodies are required by their natures to move in specific ways was replaced by the assumption that any body may move in any way, its actual path being determined by the forces exerted on it. This proved to be the true generalization.

Similarly, it was assumed at one time that nature has specific likes and dislikes; for instance, that she abhors a vacuum. This could be translated into contemporary language as the specific assumption that a law of physics prevents the density of matter from falling below a specific value. But it is now known that the true generalization about the density of matter is unspecific. The laws of physics permit any density, ranging from the high concentration that occurs in the white dwarf stars to the extreme tenuousness of extragalactic space.

A further illustration may be taken from more recent history the geometry of space. Until some fifty years ago it was assumed that this was required by the laws of physics to be of the kind known as Euclidean. If the assumption was not recognized as a specific one, it was only because it was not recognized as an assumption at all; it was thought of as a self-evident truth. Nevertheless, mathematicians had already shown that other geometries were logically possible. But very few persons believed that they were also physically possible.

Einstein, however, was prepared to take the same view about the geometry of space that the astronomer in our little story took about the number of planets in solar systems. Knowing of no law to preclude non- Euclidean geometries, he made the minimum assumption, namely that they can occur. The unifying and explanatory power of this assumption has proved to be enormous.

Yet another, rather simple, example is provided by the periodic table of the elements. The basic feature by which the chemical properties of an element are determined is the number of unit charges on the nuclei of its atoms. For the elements to be found in nature the maximum number of such charges is ninety-two, a stability limit. This limit provided a logical reason why no nucleus could be observed in nature with more than ninety-two unit charges, for a greater number would be inconsistent with the definition of a stable (or at least relatively stable) particle. But there was no logical reason why a smaller number should not occur. The minimum assumption is that a stable nucleus may carry any number of unit charges up to ninety-two.

There was a time, not so very long ago, when observation had nearly, but not quite, justified this assumption. Nearly all the numbers had been observed, but there were a few gaps. One of these was seventy-two. Another was zero.

Those who subscribed to the doctrine that a scientist must never, never believe what he cannot observe would have been precluded from believing

21

that nuclei with seventy-two unit charges could occur. But few, if any, scientists carried their faith in empiricism thus far. Their faith in the Principle of Minimum Assumption was the stronger, at least about the number seventy-two. So an element corresponding to this number was predicted purely on the basis of this principle. Its properties were also inferred and predicted. The event justified faith in the principle, for the time came when an element with seventy-two unit charges was observed and found to have the predicted properties. It received the name 'Hafnium'. Had faith in the Principle of Minimum Assumption been a little stronger and the word 'any' taken a little more literally, scientists would also have predicted nuclei, or at least particles, with zero charge. For they would have noticed that there was neither a law nor a logical reason why such particles should not occur. Having satisfied themselves about their logical possibility, they would next have worked out what properties would follow logically from the definition of a particle with zero charge. They would have decided that it could not attract any satellite electrons; that it would pass through any atom without being deflected by the charges on the nucleus of the atom; that therefore no vessel could contain it; that it would not enter into any chemical reaction. They would, in short, have predicted the neutron together with its properties. Subsequent observation of neutrons would then have served to justify the choice of a minimum assumption. Actually the sequence of events was reversed. The neutron was observed first and its occurrence and properties were explained afterwards. But no other hypothesis was needed for the explanation than that any particle may carry any number of unit charges, including zero. The neutron provides one of many illustrations of the great explanatory power of the Principle of Minimum Assumption. Here we have one of the great missed opportunities in science.

Let me quote one further illustration of the power of this principle. It is the discovery of the positron by Dirac. The scientific work that led him to predict it is recondite and need not be described in detail. A few salient facts, deliberately presented in an over-simplified form, will suffice to point a moral.

One of Dirac's equations had two solutions, as happens when one solves a quadratic equation. One of the terms in both these solutions represented energy; but it occurred with the positive sign in one of them and with the negative sign in the other. The positive sign caused no difficulty. It represents energy as we know it. But the negative sign could mean only that the solution applied to a system that contained negative energy. It was difficult to give meaning to negative energy; but this was not the only objection to the second solution. Just as, at one time, no one had observed particles that had seventy-two or zero charges, so no one had observed a state of negative energy.

Had Dirac been a slave to the doctrine that what is not observable has no place in reality he would have had to assume that a specific law of physics prohibits a state of negative energy. But instead of assuming this he allowed himself to be guided by the Principle of Minimum Assumption. He saw that it would involve a specific assumption to deny the possibility of negative energy and a minimum assumption to accept that possibility. His reasoning was strictly analogous to that of the astronomer who finds it more consistent with scientific method to postulate than to deny that some solar systems have thirteen planets.

Dirac's next step was to seek possible reasons why negative energy had never been observed. He rejected the facile answer that there was no such thing. Having satisfied himself that negative energy was logically possible he was convinced that it was also physically possible. The answer that Dirac did find need not concern us, but what is relevant is that in seeking it he reached the conclusion that evidence for negative energy would be provided by a particle with the mass of an electron and positive unit charge. This conclusion did not involve any additional hypothesis or assumption; it was a logical inference from work based on no other hypothesis than that the minimum assumption is always the true generalization.

That the predicted particle had no more been observed than a state of negative energy did not shake Dirac's faith in the principle, a faith that was justified when, some years later, the particle was actually observed. It is called the 'positron'.

Until this happened there were doubts about Dirac's prediction; and it is important to appreciate their nature. It was thought that the reasoning might have been faulty; that some essential fact might have been over- looked; that the mathematics might have contained an undetected error. The eventual discovery of the positron served to allay doubts of this kind. But they were all doubts as to whether a state of negative energy was really logically possible. Few doubted that, if it was, it was also physically possible and would occur occasionally.

Examples where, in physics, the Principle of Minimum Assumption has led to new and valuable discoveries, when it has served to predict, to explain, to unify could be multiplied indefinitely. When a minimum assump- tion has been used as a basis of the subsequent reasoning, a number of valuable conclusions have followed from it by a process of logical inference and without the need for any additional hypotheses. But I cannot recall a single instance in physics where a specific assumption has, after scrutiny, been maintained as a true generalization. For such examples one has to turn to history, to biology, to the social sciences, to the study, in other words, of systems that come under the influence of life.

2.4: *The Concept of a Cosmic Statute Book*
I have introduced the expression Cosmic Statute Book into Chapter I and have

discussed this concept in detail elsewhere. [I, II] It must suffice to mention here only one or two points connected with it.

Laws are always concerned with generalizations. The laws that govern the formation of companies, finance acts, the rule of the road, apply to all companies, to all tax payers, to all road users. This holds equally for what are called the laws of physics. But there the resemblance ends.

Laws imposed by authority can be, and often are, recorded in statute books, so I shall say that they are of the statute book kind. What charac- terizes them is that they demand a specific choice between alternatives all of which are logically possible. They require, for instance, that traffic shall keep to a certain side of the road and prohibit passage on the opposite side.

In the formulation of such laws the words 'any' and 'either' must not occur. If they did the law would be meaningless. Further, such laws do not include what is the logical consequence of other accepted principles. If they did they would be redundant. Thus no country would enact a law to say that people may drive either on the right or on the left. Such a law would enforce nothing and prohibit nothing. Nor would any country enact a law to require that two twos shall be four. It would be so whether the law were enacted or not.

The question now arises whether there are laws of physics that prohibit anything that is logically possible. We believe that it is logically possible for a solar system to have thirteen planets, and we know that it is logically possible for a particle to have seventy-two and zero unit charges, for space to have a variety of different geometries, for energy to occur in the negative state. If, nevertheless, there was a law of physics that prevented any of these possibilities from being realized it would be of the statute book kind. Such a law would have no place in a unified physics. It could not be in- ferred from any known principle. It could not be explained. It could be discovered only by observation.

According to the Principle of Minimum Assumption there are no such laws in physics and I have shown above with the help of a few examples that physicists often act on the belief that it is so. Their belief can be expressed by rewording the Principle of Minimum Assumption as follows: In physics a generalization that is logically possible is also physically possible. It can therefore be represented by an actual example and is so represented with a frequency that is determined by statistical considerations only.

[I] R. O. Kapp, *Science versus Materialism,* Methuen, 1940, Chapters XXII and XXV.
[II] R. o. Kapp, *Facts and Faith,* Oxford University Press, 1955.

Yet another formulation of the same principle is as follows: For the physicist there is no such thing as a Cosmic Statute Book.

This negative formulation brings the principle into the category of what Sir Edmund Whittaker called 'postulates of impotence'. It tells us what one cannot do. It says that one cannot base a true generalization in physics on a specific assumption. Therewith it has a faint resemblance to a formulation of the principle of conservation of energy that says: 'One cannot make a perpetual motion machine.' This negative wording has its advantages sometimes. In mechanics one tests conclusions for their conformity to the principle of conservation of energy and one may reject them by saying: 'That is equivalent to inventing perpetual motion.' If we got into the habit of testing our conclusions for their conformity to the Principle of Minimum Assumption, we should similarly find ourselves saying sometimes: 'That is equivalent to an entry in a cosmic statute book.'

However, the Principle of Minimum Assumption is far from having reached universal acceptance. People have not become very articulate about it. It is by no means applied with the uncompromising consistency that it needs. Those who do apply it do so instinctively rather than deliberately and many of them would oppose my plea for elevating the rule of economy of hypotheses to the status of a great principle of physics. Nevertheless, this is exactly what I shall do in the ensuing pages. On the more superficial view this book might be regarded as an ingenious explana- tion of certain cosmic phenomena. But what I have to say was not found as a result of a search for explanations. It was found as a result of exploring some implications of the Principle of Minimum Assumption and I am hoping that a more profound view will be taken of what I have to say than that it constitutes some ad hoc explanations. I should like it to be regarded as an example of the way in which the Principle of Minimum Assumption can be justified by its unifying and explanatory power.

PART II

THE PAST AND FUTURE DURATION OF MATTER

Chapter 3

Hypotheses About the Origin of Matter

3.1: *A List of Possible Hypotheses*
As I have shown elsewhere, [III] all hypotheses about the duration of matter that anyone has so far been able to put forward comprise a combination of one of those in the following (A) list with one in the (B) list:

There are in all nine possible combinations. Three of these are obtained by combining respectively (Al) with (Bl), (A2) with (B2), and (A3) with B3). They

(A)
Hypotheses about the Duration of Matter and Energy in the Past
(Al) All matter and energy have existed for all time.
(A2) All matter and energy have existed for approximately the same length of time, *i.e.* from the date of the Creation. According to this hypothesis no particle has existed for a longer time than has elapsed since the Creation began nor for a shorter time than has elapsed since the process of Creation was completed.
(A3) Any particle of matter or quantum of energy may have existed for any length of time. As will be shown later, this is a way of saying that matter and energy are originating without cause, continuously, at random, and not as a result of anything in the existing state of affairs.

(B)
Hypotheses about the Duration of Matter and Energy in the Future
(Bl) All matter and energy will continue to exist for all time.
(B2) All matter and energy will continue to exist for approximately the same length of time, which is the time that will elapse until the End of the World. According to this hypothesis no particle will last for a shorter time than will elapse up to the beginning of the End of the World nor for a longer time than will elapse until the process of destruction is completed.
(B3) Any particle of matter or quantum of energy may cease to exist at any time. As will be shown later, this is a way of saying that matter and energy are disappearing without cause, continuously and by extinction, at random, and not as a result of anything in the existing state of affairs.

[III] *The British Journal for the Philosophy of Science,* November, 1955, Volume 6, No 23 are symmetrical. The remaining six combinations are all asymmetrical. As every one of the combinations is just as much an hypothesis as any other, each has to be assessed by the criteria that are applied to hypotheses when a choice has to be made between them. One of these is the criterion of minimum assumption, Occam's razor. Let us first assess the various possible hypotheses about the past and future duration of . matter by this criterion. Assessment by other criteria is, of course, also necessary and will follow later.

3.2: *The Hypothesis of Continuous Existence in the Past*

In this chapter the hypotheses in the (A) list only will be discussed. Those in the (B) list will be discussed in Chapter IV.

(Al) fails badly by the criterion of minimum assumption. As will be shown below, it can be maintained only with the help of a number of assumptions that are far from minimum ones.

A minimum assumption is recognized, it will be remembered, by use of the word 'any' in its formulation. The statement that all matter and energy have existed for all time precludes the use of this word and would seem to bring (Al) into conflict with the criterion of minimum assumption. But this is not the major objection. I should feel diffident about applying the criterion thus rigorously. In spite of verbal appearances it is just possible that 'all time' is really less of an assumption than 'any time'. So I should hesitate to dismiss (Al) simply because its formulation does not contain the word 'any'.

Irreversible Processes: The true objection is deeper and stronger. It is that many other very specific assumptions are implicit in (Al). This becomes apparent as soon as one tries to reconcile this hypothesis with a simple fact of observation. This is that the world we live in is a changing one. Let me show why this presents us with a serious difficulty as soon as we try to support (Al).

Certain observed processes are irreversible. The most familiar of them is the process of falling. Things fall from a region of higher to one of lower potential and not vice versa', in colloquial terms things fall downwards and not upwards. One can describe the process as that of following a potential gradient and its result is such that the total potential energy is reduced by the process, it being converted into kinetic energy. As is well known, the conversion of potential into kinetic energy is equivalent to an increase in entropy and is a manifestation of the second law of thermo-dynamics.

Another manifestation of the same law is the equipartition of energy between colliding particles. When a particle that is at rest, but free to move, is hit by a moving particle, the one at rest acquires kinetic energy at the expense of the one that hits it. Again, when two particles of irregular shape collide some of their kinetic energy of rectilinear motion is translated into that of spin. It is easy to deduce from these facts that, given sufficient collisions between

28

moving particles, there will be a tendency for those that began with little kinetic energy to acquire more and for those that began with much to lose some of it. Eventually, therefore, the average kinetic energy in any sample large enough to be typical will be the same as in any other. The temperature of a body is proportional to the average kinetic energy of its component particles. Two substances in which the particles have different average kinetic energies are therefore at different temperatures. When they are brought into contact the kinetic energies come to be shared equally and the temperature difference is smoothed out. This is but another way of saying that heat flows by conduction from hotter to cooler bodies and not vice versa.

There are many other irreversible processes. Electric charges tend to neutralize each other. Chemical reactions tend to be such that the resultant compounds are less reactionable than the substances from which they were formed. Exothermic reactions, in which heat is liberated, are more probable and frequent than endothermic ones, in which it is absorbed; and so forth.

Such irreversible processes tend towards a terminal condition. If things keep on falling, the time must come eventually when everything that can fall will have done so until it can fall no further. If bodies in motion keep on interacting, the time must come when the total kinetic energy is shared equally between all of them. If heat keeps on flowing from hotter to cooler bodies, temperatures throughout the universe must eventually be equalized; the world must reach the condition called by German philosophers the 'warmetod' (heat death). Similarly all electric charges must be neutralized and all chemical compounds reach a state in which they cannot act chemically on each other.

All this can be expressed as a principle that I propose to call the Principle of Stabilization. By this principle processes in a self-contained system tend to reduce their causes. If such processes continue for long enough all causes must be reduced so much that they cease to operate. When this has happened there can be no more change.

But we do observe change around us. Things do continue to fall. The ponderable contents of the universe have by no means reached positions where they can fall no further. About one-half of these contents exist in the form of a thinly diffused gas in interstellar space. All parts of this gas are in some gravitational field, for such a field surrounds every star and stretches into the outer distance, where, though very faint, it exerts a finite pull on every molecule of the diffuse interstellar gas. So this gas must be falling very slowly towards the nearest stars. The terminal condition, in which all of the interstellar gas has completed its falling, is still a long way off.

Similarly, great differences of temperature are still observed. Heat in large amounts is flowing from hot to cool surfaces. Interacting bodies are still far

29

from sharing their kinetic energy equally. Particles in collision still impart considerable increases in spin to each other. Electrical and magnetic potential gradients abound. Highly reactionable chemical substances are found throughout the universe. In short, what makes for change is to be found in abundance. The hypothesis of continuous existence in the past (Al) fails to explain this.

(A.I) Can be Saved only by Piling Hypothesis upon Hypothesis: When an hypothesis fails to explain observed facts one has the choice between two alternative courses. The first is to abandon the hypothesis in favour of another one with superior explanatory power. The second is to retain the hypothesis and to combine it with further ones in such a way that the combination does explain the observed facts. One could, for instance, attempt to reconcile the hypothesis of infinite past duration of matter with the observation of present change by adding sundry additional hypotheses about the laws of physics.

A collection of carefully reasoned papers on the age of the universe has appeared in *The British Journal for the Philosophy of Science.* [IV] There, several of the additional hypotheses are mentioned that might serve to save (Al) and they are critically discussed. I do not want to cover more of the same ground here than is absolutely necessary, but I do want to emphasize that all attempts to save (Al) have to be rather desperate and cannot be reconciled with the way a physicist regards the material universe.

It could be thought that, perhaps, those processes that seem to us to be irreversible are really cyclical. The expansion of space is one of these; it has actually been suggested that it may alternate with contraction and be analogous to the motion of a pendulum, which slows down as it approaches the limit of its swing and then reverses. In the above-mentioned collection of papers J. T. Davies says 'it is impossible to say whether an expansion might not be merely a phase of a more complex phenomenon such as pulsation'. Ernst J. Opik speaks of 'the collapsing Universe rebounding from the elastic forces of the nuclear fluid at a state of maximum compression'. On this hypothesis the moment when any given irreversible process began is not a starting point but only a turning point.

This hypothesis can sound plausible when it is applied to the expansion of space. But this expansion is only one of a great number of processes that are found to be irreversible. If we are to develop a sound cosmology we dare not overlook any of the others, and when it is applied to other processes the hypothesis of a pulsating universe can sound plausible only so long as one thinks of these processes in a vague and abstract way. If we think of entropy only as an algebraic symbol in a mathematical expression, we shall find no difficulty in believing that in a self-contained system its rate of change may

---

[IV] *The British Journal for the Philosophy of Science,* November, 1954, Vol. 5, No. 19.

sometimes be positive and sometimes negative. Perhaps, we might then say, the second law of thermo-dynamics defines cyclical and not unidirectional processes; perhaps during infinite time a tendency for the entropy of a self-contained system to increase alternates with a tendency for it to decrease. Perhaps a tendency for processes to reduce their causes alternates with a tendency for them to increase their causes.

But when one thinks of the concrete reality that finds its generalization in the second law, those 'perhapses' fade to an unconvincing pallor. The law expresses, among other things, the fact that colliding bodies tend with time to share their kinetic energies equally. Perhaps, one is led to assume, there have been times when a body that was at rest and was hit by a moving one imparted kinetic energy to the body that hit it and lost some of its own (non-existent) kinetic energy. Perhaps spinning bodies once-upon-a-time gave up their spins on collision and converted their spin energies into energy of rectilinear motion. Perhaps at one time gravitation was reversed and things fell up instead of down. Perhaps inert chemical substances used to act on each other and cinders converted themselves into coal. Perhaps heat used to flow from the cooler to the hotter surface. Perhaps electrical potential gradients used to grow steeper when charges were neutralized. Perhaps the laws of physics are not universal but change from time to time. Perhaps the Cosmic Statute Book comes up for periodic revision. Perhaps the last occasion was the precise moment when a pulsating universe had reached its maximum concentration and began to expand. Perhaps all the present irreversible processes started together just then. If we attempt to save (A1) we shall run the risk of drowning in an ocean of strange hypotheses.

Most, if not all, of them are untenable anyhow. That objects of irregular shape tend to spin when they collide is not the consequence of a transcendental law but of simple mechanics. It can be easily shown that there are innumerable ways of colliding that cause spin for one way that does not. The laws of physics allow the objects to collide in 'any' way and so nearly all collisions are of the spin-causing kind. The irreversible process by which rectilinear motion tends to be converted to spin does not arise from a specific law but from the absence of a specific law. To say that 'perhaps' it was the other way about at one time is to say that perhaps there was a specific law at that time to control collisions, a law that has since been struck off the Statute Book.

It can, moreover, be shown that many of the so-called laws that seem to show processes to be irreversible are merely true by definition. One defines a field of force in terms of the direction in which things fall in it. To say that they fall in the opposite direction is to contradict the definition. One defines temperature gradient similarly in terms of the direction in which heat flows. To say that it flows from the hotter to the cooler surface is implicit in the definition. This is why, if we seek to offend against the rule of Economy of Hypotheses in the forlorn hope of saving (A1), we shall do so in vain. There

31

would rarely be the need to reproach us for piling one hypothesis on another. One could refute us by logic.

For the foregoing reasons the notion that the whole existing contents of the material universe have existed for all time seems, today, to have but few sponsors among serious scientists, whether the notion be of a universe in a process of unidirectional change or of pulsation. It fails badly by any acceptable scientific criterion and certainly by that of minimum assumption.

3.3: *The Hypothesis of Past Finite Existence*
One can explain the fact that there is still change in the universe if one adopts the hypothesis that all the contents of the physical universe have existed for a definable time. This is (A2), and at the time of writing it has considerable vogue. Sundry estimates have even been made concerning the age of the universe.

Irreversible Processes have been Claimed to Provide Evidence for (A2): For this purpose a few of the observed irreversible processes have been used as clocks. It is assumed that the universe started in some specific condition and that it has been departing at a known rate from this condition ever since. The degree of departure is taken as a measure of the time that has been occupied by the process.

One of the clocks is provided by the expansion of space. It is assumed by supporters of (A2) that the whole contents of the universe started in a small volume of very high mass density and have been spreading out more and more ever since. By measuring the rate of expansion and the distance between discrete objects one arrives at the time assumed to have passed since the expansion began. One of the other clocks is the ratio of helium to hydrogen in stars; it is assumed, with considerable justification, that all ponderable matter has begun its existence in the form of hydrogen and that its slow conversion into helium is among the irreversible processes that give a reliable time scale. Another clock is provided by the partition of kinetic energy between stars, and yet another one by the proportion of certain isotopes found in rocks. These and several other clocks give roughly the same age, namely between four and seven thousand million years.

I do not propose to discuss in detail the observations on which the various estimates are based. They have been fully described by others and there is no reason to doubt that the estimated time-interval of some thousands of millions of years has a cosmic significance of some sort. It does not therefore seem likely that hypothesis (A2) will have to be rejected on the grounds that the various clocks provided by irreversible processes are unreliable; the agreement between them is too good for their readings to be lightly dismissed.

But we must take care to draw the correct conclusions from those readings.

One may safely conclude that since the clocks were, so to speak, wound up the part of the material universe where we are has been behaving as it does today. But this conclusion is equally compatible with (A1), with (A2), and with (A3), as has been pointed out by sundry writers of the papers that appeared in the above-mentioned 'Age of the Universe' number of *The British Journal for the Philosophy of Science*.

Supporters of the hypothesis that a pulsating universe has been existing for infinite past time, (A1), could point out that the moment when the clocks started need not be interpreted as a moment of origin; it could equally well be a moment of change in a universe that had existed in a different form before then and had obeyed a different set of laws. The strange collection of hypotheses that would have to be designed in order to save (A1), and that has been mentioned already, may be refuted by various arguments, but it is only fair to add that it cannot be refuted by the evidence of the clocks. Michael Scriven, one of the contributors to the papers on the age of the universe referred to above, speaks of 'phases in the Universe prior to the present "expansion",' as a description of a possible view, though not one to which he seems personally to commit himself. He also asks the pertinent question: 'What is the difference between saying that the Universe has an origin in time and saying that each of the processes of change within the Universe has an origin in time ?' The implication is, of course, that the evidence of the clocks might equally well be used as evidence for the pulsating universe that, according to (A1), would have existed for all time.

It will later become apparent that the evidence of the clocks could also be used to support (A3), the hypothesis of continuous origin. But that can be easily understood only after (A3) has been discussed. Suffice it at the present moment to refer again to the above-mentioned papers, where J. T. Davies shows this cogently when he says 'thus the theory of continuous creation provides us with a picture rather like that of a living population; individuals are born, age and die, but the population may remain in a steady state'. What, in other words, the clocks record is not necessarily the age of the universe as a whole. Each clock may only record the age of the particular part of the universe in which it is placed.

It will appear in due course that one should expect matter that was originating continuously, in accordance with (A3), periodically to form nebulae, like our galaxy. One should expect the various clocks placed in any one of them to record its age and to give consistent readings between themselves.

In short, claims that there is observational evidence for (A2) have, to say the least, been greatly exaggerated. It is doubtful whether there is any evidence at all for it that is not also evidence for either of the other alternatives.

(A2) Fails by the Criterion of Economy of Assumption: Perhaps more scientific considerations will be found some day in favour of the hypothesis of a definable time for the past existence of matter than have been' found hitherto, but while we wait for these we must also assess the hypothesis by the criterion of minimum assumption. By this it fails at least as badly as the hypothesis of past infinite existence. This must cause disquiet to any scientist who is contemplating the hypothesis. To emphasize this disquiet I have on a previous occasion referred to (A2) as the 'once-upon-a-time theory'. [v] My reason was that, like a fairy-tale, it fails conspicuously to conform to Occam's razor.

(A2) postulates, for instance, two (and not only one) significant and quite specific moments in all time. The first is the moment at which it has to be assumed that the very first element of the universe came into existence. It is the moment before which it is assumed that there was nothing, The second specific and significant moment of time is the one at which the very last element of the material universe came into its correct position. It is the moment before which is assumed that the universe was incomplete, and after which it is assumed that it was complete. These two moments, it is assumed, have never been repeated in the whole of time, Two more-specific assumptions are hard to imagine.

It is alien to the spirit of physics to contemplate unrepeated and unrepeatable events. It is part of the physicist's faith that events are all repeatable whenever similar conditions are reproduced, and that given sufficient time similar conditions will always be reproduced. To postulate two specific moments in time is to reject this. Physicists will be unwilling to do so without a great deal of evidence.

But this is not even the whole of the objection. Hypothesis (A2) implies that the laws of physics, as we know them, did not come into operation until the second of the two significant moments. The implication of the hypothesis is that there was once in the past a period of finite duration between those two moments, which may have lasted for seconds only , or for millions of years, but was in any case finite. From a specific moment in the year t_1 B.C. until another specific moment in the year t_2 B.C. it is implied that the laws of conservation of energy and of matter were suspended, that they came into force precisely at the second of the two moments, and that they never can be suspended again.

For all that can be proved to the contrary it may have been so. But an hypothesis that implies it cannot be called a minimum one. The belief that the laws of physics hold, and have held, and will hold, everywhere and at all times is a strong one. This is why it demands a substantial intellectual effort to be persuaded of (A2). But the effort has not to be directed at understanding

[v] R. O. Kapp, *Science versus Materialism,* London, 1940, Chapter XXIV

something; it has to be directed at ignoring something, namely the implications of the hypothesis. The period between the beginning of the Creation and the completion of the universe is one that does not bear thinking about. To make (A2) acceptable one must prevent this period from intruding on one's attention. An hypothesis that calls for this ignoble kind of intellectual effort must be suspect. That it fails conspicuously by the criterion known as Occam's razor would not alone be a sufficient reason for rejection of (A2). But it seems also to fail by all other scientific criteria. The evidence for it is slight and of doubtful validity. It has but little unifying and explanatory power. If it is to be justified at all, it can only be on non-scientific grounds.

3.4: *The Hypothesis of Continuous Origin*

I doubt whether the hypothesis of continuous origin (A3) would ever have suggested itself if both the other two hypotheses had not proved so unsatisfactory. It is fair to say that I did not arrive at (A3) by a direct approach to the problem but by a process of elimination consisting of the rejection of (A1) and (A2). To be justified, (A3) will have to be free from the glaring defects of the other two hypotheses. It will have to meet the Principle of Minimum Assumption adequately, it will have to manifest greater explanatory power than the other two, and it will need better support from the evidence of observed facts.

The justification of hypothesis (A3) on the grounds that it involves a minimum assumption was presented when I first advocated this hypothesis in 1940. To emphasize this justification I called it the 'at-any-time' theory. Others have advocated the same hypothesis since and found sundry other reasons for its justification. [VI, VII, VIII] That it deserves serious consideration is therefore by no means a new claim. Its great explanatory and unifying power will appear in subsequent pages, together with the evidence that supports it. For the moment I shall simply discuss its main features in a little detail.

What is it that Originates? To the claim that matter is originating continuously it is natural to ask In what form?' At the present stage of knowledge concerning the basic constitution of the material universe it is not possible to answer this question with any precision. One would naturally not assume that complex structures come into spontaneous existence completely synthesized. To do so would be to assume that the process of originating was

---

[VI] V. Hoyle, Stellar Evolution and the Expanding Universe, *Nature*, 1949, 163, 196.

[VII] H. Bondi and T. Gold. The Steady State Theory of the Expanding Universe. *Monthly Notes Royal Astronomical Society*, 1948, 108, 252.

[VIII] W. H. McCrea, The Steady State Theory of the Expanding Universe, *Endeavour*, 1,1950, 9, 3; Relativity Theory and the Creation of Matter, *Proceedings of the Royal Society*, 1951,206,562.

also coupled with a process of ordering to specific forms and that would not conform to the Principle of Minimum Assumption. For this reason none of the advocates of (A3) has suggested that complete molecules of chemical compounds are among the originating bodies. Nor has it been suggested that the nuclei of any of the heavier elements are among them. These are known to be composed of particles that are more elementary than they, such as protons and neutrons, and the minimum assumption is that the originating particles are the most elementary of all. It is, however, impossible to say in the present state of knowledge what the uttermost elementary particles are.

If all so-called elementary particles had the same mass one might regard them as truly elementary. But indivisible particles are known with various inertial masses. It may be that each of them is in some way composite and that the respective masses of the electron, the proton and the sundry mesons are each the consequence of a synthesis from some even more elementary components. One should hesitate to regard any of these particles as the uttermost elementary components of the material universe.

The same objection does not apply to unit charge, which has always precisely the same value on the electron with the negative sign and on the proton with the positive sign. There is, however, convertibility between various elementary particles, such as between neutrons, protons, electrons, positrons and others. A proton may, for instance, convert into a neutron, a positron and a neutrino. One is thereby led to doubt whether unit charge is a quite basic constituent of matter.

It is doubtful even whether the most basic constituents ought to be called particles. For there is also convertibility between particles and photons, as there is between mass and energy. In the sense in which protons and neutrons are common constituents of atomic nuclei some as yet undefined components of matter may be common constituents of positive and negative charges, elementary masses, and photons. If so, these undefined components will be no more like the things of which they are components than neutrons and protons are like the nuclei into which they synthesize.

Bearing in mind that matter is not so much *in* space as *of* space, the most accurate description, and a non-committal one, of the uttermost. elementary constitutent of matter may be *a bit of differentiated space*. But the inseparability of matter and space has not yet become a familiar concept, so I propose to speak here, even more non-committally, of the continuous origin of *elementary components* of the material universe.

That I have to speak thus non-committally proves that the hypothesis of continuous origin (A3) is incomplete in the form in which I have been able to state it. In the next chapter I shall develop the hypothesis of the continuous extinction of matter and this will be equally incomplete. I have no answer to the important question: 'What is the basic unit that originates and what is the basic unit that becomes extinct?' The question abounds in difficulties. Some of these will be discussed in Appendices G and H.

One of the conclusions to be reached in Appendix H is worth mentioning now, for it will help towards understanding what is said in the main parts of this book. The conclusion to which more than one line of reasoning leads is that a complete atomic nucleus, together with its electric charges, is a basic unit and becomes extinct as a whole.

This conclusion is important. If a part of a nucleus were to become extinct while the remainder continued to exist, this remainder would not only be radio-active: it would also be the nucleus of another element. But observation shows that it is not so. Hence the hypothesis of continuous extinction would have been severely shaken if the line of reasoning followed in Appendix H had led to the conclusion that only a component part of a nucleus becomes extinct at any particular moment.

Even if the suggestions put forward tentatively in Appendix H are corroborated the hypothesis will, however, still be incomplete. Many difficult questions will remain unanswered. In drawing attention to this. I should also draw attention to what I have said on page 16 of the Preface. To say that a theory is incomplete is not the same as to say that it has been falsified.

Chapter 4

Hypotheses About the Disappearance
of Matter

4.1: *The Alternatives*

To obtain a complete hypothesis about the duration of the contents of the material universe one must combine one of the three hypotheses in the (A) list at the beginning of Chapter 3 with one of the three in the (B) list. The (A) list, it will be remembered, is concerned with the past and the (B) list with the future duration of matter. Of the nine possible combinations eight have to be eliminated.

The elimination cannot be based on scientific proof, for it does not appear that any one of the nine combinations can be proved false. But if one abandons the criterion of proof and adopts instead the criteria of minimum assumption, maximum explanatory and unifying power and satisfactory evidence, most of the combinations can be eliminated without much difficulty. It has already been shown that (A1) and (A2) fail badly by these criteria and I shall have several further occasions to confirm their inadequacy. So there remains only (A3) in the (A) list.

Of those in the (B) list it will be convenient to consider (B2) first, for this is the one that it is easiest to eliminate. (B2) is the hypothesis that the whole contents of the material universe will have ceased to exist after a definable time in the future. Apart from having the doubtful virtue of giving a very literal interpretation to the theological doctrine of the End of the World it has nothing to recommend it.

Consider its implications. Unless the further strange hypothesis is added that a process can be completed in zero time, supporters of (B2) must assume that a moment will arrive during the future year T_1 A.D., when the extinction of the contents of the material universe will begin and they must assume a later moment during the year T_2, when the process of extinction will be complete. Perhaps the hypothesis that a process can be completed in zero time would find supporters among laymen. But physicists do not know of any observation, experiment or line of reasoning by which to justify such an hypothesis.

If we believe that extinction is prevented in our time by certain specific physical laws, we could support (B2) only by assuming that these laws will be suspended for a period dating from the first of the two significant moments in the year T_1 and lasting until the second significant moment in the year T_2. We should, however, have to assume that the suspension will be one-sided, permitting the extinction of mass and energy but prohibiting the origin of anything new. It would be interesting to work out the implications of this hypothesis in detail. A physics from which one basic law was removed would differ in many other respects from that with which we are familiar.

From the moment when this new physics applies and matter begins to behave in new strange ways it would have to be assumed that the contents of the material universe will be dwindling until finally just one elementary component will be left. Then this, too, will disappear to mark the great transition from something to nothing.

This hypothesis can no more be disproved than its counterpart (A2), which assumes the same process in reverse for the past. But both are equally foreign to the physicist's habits of thought. The picture that they present looks more absurd the more closely it is examined; and this is just as true when the picture is relegated to a remote past as when it is relegated to a remote future. It is therefore not at all surprising that (B2) has no serious supporters among scientists, but it is surprising that (A2) has. Why, one is led to ask, should anyone who cannot believe that the laws of physics will ever be changed in the future believe that they may have been changed once upon a time in the past? Why should the notion of two specific, unrepeated and unrepeatable moments in the years T_1 and T_2 A.D. be rejected while the same notion is not even critically discussed when it is applied to the years T_1 and T_2 B.C.? Why should anyone who finds the notion of a future transition from something to nothing unacceptable be able to accept the notion of a past transition from nothing to something ? I do not know the answers to these questions, but the fact remains that attempts to justify (A2) by extrapolation from present observations have, at the time of writing, become fashionable and are being seriously discussed among scientists, while an attempt to justify (B2) by similar methods has not yet been made and would probably be deprecated.

Whatever it may be that causes (A2) to appear attractive need not, however, concern us here. (Al), (A2) and (B2) must all be dismissed by those who become aware of the tangle of additional hypotheses that are needed to preserve them. There remain thus for serious consideration only two of the nine possibilities: the combinations of (A3) with (Bl) and of (A3) with (B3). The former assumes that every elementary component of the material universe may have come into existence at any time and that a conservation law prohibits it from ever ceasing to exist. It attributes to each component a finite past and an infinite future, i.e. it assumes that one cannot trace the existence of any given component indefinitely back into the past, but that one can predict an indefinitely long future for it. (A3) coupled with (Bl) assumes a combination of past impermanence with future permanence. It can be called the Hypothesis of the Asymmetrical Impermanence of Matter. The combination of (A3) with (B3) can then be called the Hypothesis of the Symmetrical Impermanence of Matter; it assumes impermanence both for the future and for the past.

(Bl) postulates a specific law, which requires that every elementary component of the material universe shall last for all time. (B3) does not postulate this nor any equivalent law. It says that *any* component may become extinct at any moment of time, which is what would happen in the absence of a law. So (B3) seems to meet the criterion of minimum assumption better than (Bl). But I do not want to exaggerate the importance of this. I have already said that 'all time' may perhaps be less of an assumption than 'any time'. However, it will emerge from these pages that (B3) also meets the criterion of explanatory power much better. But this does not mean that (Bl) can be lightly dismissed. It is arguable that the Principle of Minimum Assumption may not always apply. (Bl) does have a little explanatory power. It is, moreover, sanctified by long tradition and has the backing of considerable authority.

This backing prevents its hypothetical nature from being easily recognized. We tend, understandably, to believe what we have been taught and to take for granted that it must have been subjected to the rigid canons of scientific proof. We are often told that good scientists, following Newton's precept, never adopt hypotheses and so we do not doubt that what has the backing of their authority must be irrefutable fact. It is a part of human nature to accept what is easy to believe and to reject what is unfamiliar. For all of these reasons many people are likely to deny hotly that the permanence of matter and energy (or at least their future permanence) is an hypothesis. They will contend that (Bl) is fact and only (B3) hypothesis.

I think that this view may well have been taken by the more recent supporters of (A3); for they have all adopted the Hypothesis of the Asymmetrical Impermanence of Matter[1] Otherwise their reason for supporting (Bl) is not easy to understand. They can hardly have thought that it was the more attractive alternative. By the unreliable criterion of attractiveness one should expect symmetrical impermanence to be the favoured hypothesis. A more probable reason is that (Bl) is so firmly established and generally accepted that (B3) did not even enter their thoughts. This surmise is confirmed by the fact that the supporters of asymmetrical impermanence have never undertaken a critical comparison of the alternatives, as they would surely have done if they had noticed that there were alternatives.

He who has a new point of view to present cannot afford to ignore contemporary opinion and so I have to attach much importance to the relative strengths with which each of the hypotheses in the (A) and (B) lists is held today. From reading and conversation I have gained the impression that support for (Al) would be very weak. Many will say that, in attempting to refute it, I am only flogging a dead horse. But support for its counterpart (Bl) is likely to be so strong that my attempt to refute it is likely to evoke a different metaphor: that of tilting at windmills.

Nevertheless, it was a combination of (A3) with (B3) that I advocated when I first published the hypothesis of continuous origin in 1940 and I propose to show here that it alone is consistent with a cosmological model that accords with the observed universe.

Before I leave this theme I should like to emphasize that the words 'origin' and 'extinction' are to be understood literally and not as synonyms for 'conversion from one form into another'. It would deny my intention to assume, for instance, that, when matter becomes extinct, something else, such as energy, must necessarily take its place.

4.2: *The Relative Rates of Origins and Extinctions*
Let us imagine that a region of space can be so isolated that no matter or energy passes its boundary. According to the Hypothesis of Asymmetrical Impermanence the content of this region must continuously increase; for new matter will originate in the region while none will leave it. According to the Hypothesis of Symmetrical Impermanence, on the other hand, some matter will be originating in the region and some becoming extinct, and so the content will not necessarily increase. It will do so if the rate of origins exceeds the rate of extinctions; but the content will decrease if the rate of extinctions exceeds that of origins.

If the region is small and the density low, origins and extinctions will be only occasional and as they are both assumed to be random, they will occur at irregular intervals. The content may decrease at one moment and increase at the next. But the larger the region the more such irregularities will be smoothed out; and if a region be chosen large enough to be a fair sample of the whole material universe the rate of origin per unit volume will be the same as for the whole and so will the rate of extinctions.

McCrea has calculated the rate at which the content in mass of the universe is increasing and finds it to be about 500 atoms of hydrogen per cubic kilometre per year. According to asymmetrical impermanence this is the gross rate of origins. According to symmetrical impermanence it is the net rate given by the excess of origins over extinctions; the gross rate is greater, and may be much greater.

It is, of course, quite impossible to observe directly the annual arrival of 500 atoms of hydrogen in a cubic kilometre of space. The mass of those atoms is very minute and could not be measured even if they could be assembled at one time and place. But there is an indirect method for estimating the rate. This is derived from the fact that, as relativity theory shows, matter and space are

41

inseparable. According to relativity theory, [IX] space is not regarded as a container but as a constituent of the material universe. To speak of the origin of matter is the same as to speak of the origin of space, and if the *content* of the material universe is increasing its *extent* must be doing so too! [X] The origin of new matter must therefore be associated with the expansion of space and the rate at which the extent of the material universe is increasing becomes a measure of the rate of increase of its content. This is fortunate. For the rate of increase of extent is easier to measure than that of content. It is done with the help of telescopes, and these reveal regions of space large enough to show an easily measured effect.

As is well known, this effect is manifest as a shift towards the red of the spectrum lines of very remote objects, called the Doppler effect. A simple calculation shows that the amount of red shift is proportional to the velocity with which a remote object is receding. Observation shows that the amount of the red shift is proportional to the distance of the object. From these two facts it follows that the velocity with which every object is receding is proportional to its distance from us. This is what one would find in a universe that was expanding uniformly and so the expansion of space is inferred from observation and known facts and without the use of additional hypotheses. The rate of expansion is expressed by the simple formula $v=Hx$, where v is the velocity of recession, x the distance of the observed star and H a constant known as Hubble's Constant. It represents the increase of velocity of recession with distance, dv/dx, observed for every object in a uniformly expanding universe and is given as 145 km/sec per megaparsec by Opik. [XI] A more recent estimate is given by Humason, Mayall and Sandage. [XII] This is based on the measurement of red shifts. for nearly 1000 extragalactic nebulae and is 180 km/sec per megaparsec.

These most careful measurements show that the rate of expansion is nearly, if not quite, constant for all regions of space-time. But they do seem to show some small lack of uniformity. There may have been a change during the long period of time that the light from the most distant nebulae takes to complete the journey that ends in the observers' telescopes. However, it will be shown later that there are good reasons for the view that the change is not between different periods of time but between different regions of space, and that it results from unevenness in the distribution of ponderable matter.

The constancy of the expansion rate, and with it the rate of increase of the

---

[IX] *See footnotes 4,5, 6 to Chapter 3.*
[X] I doubt whether this is implicit in relativity theory, but it has been assumed by others besides myself. Its justification will be found in Appendix H.
[XI] *The British Journal for the Philosophy of Science,* November, 1954, Vol. 5, No. 19, 210.
[XII] *Astronomical Journal,* Vol. 61, April, 1956, pp. 97-162.

content of the material universe, reminds us of the constant half-lives of radioactive isotopes. The reason why the disintegration of an atom of radium is regarded by physicists as an uncaused event is just this constancy, as will be explained in the next chapter. If the half-life of radium varied, one would look for a circumstance with which the disintegration was associated and expect to find a causal relation between such circumstances and the disintegration. But the fact that the rate never changes is taken as negative evidence that the disintegration is uncaused.

The same reasoning applies to the increase in the content of the material universe. If one regards the unvarying value of Hubble's Constant as evidence of a constant net increase in the quantity of matter in the universe, one will also regard it as observational support for the conclusion that origins are random, without cause, and not associated with anything in the existing state of affairs. If one accepts (B3), one will then have to take the same view about extinctions. But this aspect of Symmetrical Impermanence will be discussed more fully in the next chapter.

Chapter 5

Causation and the Judgement
of Common Sense

5.1: *Need all Events have a Cause?*
A comparison of the various conceivable hypotheses about the past and future duration of matter requires discussion of the bearing of each on the concept of causation. It is, in particular, desirable to consider what views on this subject are in accordance, respectively, with the judgement of common sense and with scientific thought. The former has, admittedly, proved fallible time and time again. But one cannot blind oneself to its potency in influencing opinion.

One of the judgements of common sense is that every event must have a cause. It is a judgement that seems to be justified by everyday experience. We observe that the tides are caused by the joint action of the moon's presence in the sky and the earth's rotation; without this combination of circumstances there would be no tides. Examples could be multiplied indefinitely. There is obviously a sense in which many events have a cause. The question is only whether we are right in assuming that *all* do.

One cannot avoid this rather awkward question by saying that it depends on what one means by the word 'cause'. For the question is more than verbal or formal. True, it is possible for different people to attach different meanings to the notion of causation. But they all mean at least that a given event, P, can only occur when there is a given circumstance or combination of circumstances, Q. Even when there is doubt as to whether P is the cause of Q or Q the cause of .P, some association between P and Q is necessary for a causal relationship to occur. When one speaks of a *causal* relationship between events one may argue about the meaning of the word 'causal', but one implies at least that there is a *relationship*.

Such a relationship is observable when there is a conversion of something into something else. There is such a relationship between water and steam when the one is converted into the other; without the pre-existence of the water there would be no steam. The same holds for other conversions; for that of hydrogen and oxygen into water; for that of the energy bound chemically in coal into the energy manifest in the heat of the product of combustion; for the conversion of some of the inertial mass in a radio-active nucleus into photons, which have no inertia and therefore travel with the velocity of light

For all such conversions, as indeed for any causal relationships, one can say that what happens depends on what is or has just been. Common sense demands that one shall be able to say this about any and every event. Does any of the three hypotheses allow one to say it about the origin of matter and energy?

The question does not arise for hypothesis (A1). To assume that the whole material universe has existed for an infinite time is to deny an origin entirely. However far back in the past one may choose to place a given moment of time, it is always, according to (A1), a moment at which the whole material universe was in existence and had been so for an infinite time. By this hypothesis one can never define an event that could be called an origin; one can never find a moment when something operated that could be called the cause of the existence of the material universe.

5.2: *Common Sense Cannot Accept the Notion of a Past Finite Existence*
One runs into a difficulty when one tries to place the origin, or creation, of the material universe at a finite time in the past. The difficulty arises, in other words, when one abandons (A1). For it is easier to assume, in conformity with (A1), that things sometimes change but can never originate, than to assume, in conformity with (A2), that once upon a time, and once only, there was a true origin.

This explains, I think, the efforts that have been made to save (A1) by supposing that the universe pulsates and that irreversible processes have periodically operated in the reverse direction. Such a view implies that there was a time, earlier than that to which the clocks of the irreversible processes point, when there already was something. It is assumed that some things were happening during that time, though we have no means of knowing what they were. Common sense can accept this. But what none of us can accept is, I believe, the notion of an infinite period of time before the beginning of the material universe during which we are required to think that there were no events and there was no activity. Philosophers, following St. Augustine, point out that it is meaningless to speak of a period of time in which nothing whatever happens.

So we find ourselves in a dilemma when we try to decide between (A1) and (A2). Are we to believe that the present material universe began during a finite period of past time without being the consequence of or related to anything that existed then ? If so, we must defy common sense and regard the origin as an uncaused event. Or are we to believe that there was already something before the moment called the beginning, not perhaps enough to be called a universe, but enough to act as the cause of the present universe ? We cannot do so. If what we assume to have preceded the present universe was changeless, eventless, inactive till the time of the origin, it might not have existed. If it was changeful it does deserve the title universe, whether it

45

resembled our own or not. The time to which the clocks point was then a moment of change and not one of origin and (Al) has to be adopted.

I have not mentioned any theological aspects of this dilemma, for to do so would be to go outside the scope of this study. But it seems to me that these are as difficult as the scientific ones and are, indeed, identical with them.

Unfortunately (A3) is as unacceptable to common sense as (A2). Both these hypotheses permit some things to have a cause but deny that *everything* has one. (A2) denies it once, and once only, for the whole of the matter and energy in the universe. (A3) denies it for the origin, as and when it occurs, of every elementary component, This has to be appreciated with uncompromising strictness. If the origin of a new elementary component were only the conversion of something that existed already into that component, if new matter and energy were but the consequence of a combination of existing circumstances, there would be no continuous origin, only continuous change. To believe this would be to support (Al) and not (A3).

A natural unwillingness to believe in the possibility of any uncaused events may explain why most supporters of (A3) speak of continuous 'creation' instead of continuous 'origin'. It is possible that, in the absence of a physical cause, they are thereby wishing to imply that there must be a non-physical one. For the term 'creation' implies a creator and would hardly be chosen by anyone who did not want to imply that. The term 'origin' on the other hand, which I used when I first put forward the hypothesis, was chosen so as to avoid this implication. I chose a term that neither asserts nor denies the theological doctrine of Creation.

5.3: *Common Sense Cannot Accept any of the Hypotheses*
If common sense must reject both (A2) and (A3), that is, both the view that there was a true origin during a finite past period of time and that there is continuous origin all the time, it must also, I am afraid, reject the notion that the whole of the matter and energy in the universe have existed for all time. For the concept of infinity has no place in the judgement of common sense. This is apparent from the colloquial meaning of the word 'infinity'. When it is used by a layman it has for him simply the connotation of a very long time; its mathematical significance eludes him.

This is important to the way the hypothesis of past infinite existence (Al) is interpreted. The true interpretation is that the universe never originated. For, as I have said already, however far back in time one may place the supposed moment of origin, the universe had already existed for an infinite time at that moment. A universe of infinite duration was as old 10^n years ago as it is today, even when n is made as large as one likes. But to the judgement of common sense the universe was 10^{11} years younger at that time.

46

The distinction between the two views of (Al) can be succinctly expressed as follows. According to what I am calling the common sense view) (Al) defines a universe that had an origin, though this was an infinite (meaning very large) number of years ago. According to the correct view, (Al) defines a universe that never had an origin.

My reason for pointing all this out is to show that in this instance the appeal to the judgement of common sense gets us no further. It rejects (Al) because this precludes a beginning and it rejects both (A2) and (A3) because they postulate a beginning. Common sense would like a beginning that is not really one; it asks for the impossible. This helps to explain the impatience that many of us feel at discussion about the past and future duration of matter. I have often felt it myself and can sympathize with those who would prefer to avoid what must seem to be no more than metaphysical argument. If I am urging that this impatience be overcome it is only because I have discovered that the discussion may lead to more fruit- ful results than one might suppose. But this can appear only later.

5.4: *Scientists can Accept (A3)*
If difficulties in the notions of causality and infinity preclude common sense from accepting any hypothesis about the past and future duration of matter, they do not seem to prevent scientists from accepting at least (A3), the hypothesis of continuous origin. For the kind of uncaused event that is implied by this hypothesis has recently become familiar to them.

Radio-activity provides an example. When an atom of radium disintegrates it emits a pulse of short-wave radiation, loses some of its substance, and turns, after a series of successive changes, into an atom of lead. What triggers the process off? Where shall one look for its cause ? Shall it be in external circumstances or in the internal structure of the atom?

If the cause were in external circumstances, the atom would disintegrate more readily when the circumstances favoured disintegration and less readily when they did not. One would then find that a lump of radium exhibited more or less radio-activity according to the circumstances in which it was placed. But what one does find is that the activity is strictly the same for a given mass of radium no matter what is done to it. The rate of disintegration of atoms in a lump of one of the isotopes of radium is always such that one half of the atoms will have undergone the process after the passage of 1620 years. There are so many atoms in a sizeable lump that, at this rate, the disintegrations do not seem to be intermittent but continuous.

One cannot reduce the rate by protecting the lump from any of the influences to which it is normally subjected and one cannot increase the rate by submitting the lump to the most violent treatment, be it by mechanical, thermal, electric or chemical means. So the cause of a disintegration cannot be among any of the detected influences to which radium is ever subjected. Those who object to the assumption that the disintegration of a radium atom has no cause might prefer to assume an undetected influence. But as the rate of radio-activity never changes, such an undetected influence would have to be assumed also to be unchanging, whatever changes might occur in its environment.

It could be suggested that the determining condition for disintegration may not lie in the outside world but in the structure of the atomic nucleus. Perhaps, it could be argued, there are certain specific differences between the nuclei of atoms of radium, and these differences become manifest in the time that elapses before the atom disintegrates. The hypothesis is that some process within the nucleus is started at the beginning of its existence, a process analogous to the winding up of the clock in a time bomb.

However, such a 'perhaps' would not meet the rule of economy of hypotheses. It would be necessary to add to the hypothesis of an undetected internal cause the further hypothesis that the causes of disintegration were nicely correlated in such a way as to ensure a perfectly even rate of activity for thousands of years. Such an addition is far from being a minimum assumption. Here the minimum assumption is that any atom of radium may disintegrate at any moment of time. If one adopts this minimum assumption one can infer that a lump of radium must have a constant half-life. The prediction that one bases on this minimum assumption is confirmed by observation and experiment. It is for this reason that scientific method precludes the complicated tangle of additional hypotheses that alone can save the common sense view of causation.

In short, the hypothesis that there is no cause suffices by itself to explain the constant rate of radio-activity, while with the hypothesis that there is a cause the constant rate takes a lot of explaining.

Physicists are well aware of all this and most of them have long ago given up the hypothesis that every event must have a cause. But laymen are often most disinclined to do so. They prefer to assemble a large collection of 'perhapses' with the sole purpose of preventing it. So it becomes relevant to look for the reason why belief in universal causation is held with such tenacity.

I think the belief lies in our anthropomorphic habits of thought. We are right in assuming causes for our own thoughts and actions. Among them are our past experience, our needs and wishes, our emotional constitution. That laws of cause and effect operate in human affairs is a fact that we observe at first hand. But we have difficulty in abstaining from the habit of projecting notions gained by introspection on to the external world. We cannot bear to think that the material universe can differ from ourselves to the extent of including among its events some that are causeless.

Be that as it may it has to be recognized that our notions about causation are often both confused and emotional. Opinion is sharply divided about the hypothesis that every event has a cause, and an attempt has been made to reconcile the views of those who are for and against the hypothesis by introducing the concept of statistical causation. It may prove a useful concept; I am not qualified to judge. But it is not relevant here. It suffices for an understanding of Symmetrical Impermanence that the causation or lack of causation that applies to the origin and extinction of matter is of the same kind as apphes to processes in atomic physics.

In the latter it is found that there is no reason why a given atom should disintegrate at a given moment rather than any other atom. Similarly Symmetrical Impermanence asserts that there is no reason why an elementary component of the material universe should originate at a given moment in a place p_o rather than in another place p_r, It also asserts that there is no reason why a given elementary component W_o, should become extinct at a given moment rather than another component W_v. To say that this lack of reason for a particular event shows the event to be governed by statistical laws rather than not governed by any laws seems to me to mark a difference in words and not in concepts.

Nothing precise is yet known about the gross rates of origin and extinction. All that can be said at the moment is that one should not expect them to be exactly equal; that would be too improbable. Further it would be contrary to the Principle of Minimum Assumption. There is also an obvious reason, based on observation, why one should expect the rate of origins to exceed that of extinctions. If it were not so, there would be no material universe to observe! The bare fact that there is one suffices, if Symmetrical Impermanence be accepted, to prove that the average rate of origins for the whole exceeds that of extinctions.

5.5: *Symmetrical Impermanence is a Derived Hypothesis*
It is important correctly to appreciate the status of the Hypothesis of Symmetrical Impermanence. (Al) and (A2), as well as (Bl) and (B2), are hypotheses in their own right. They are not inferred from anything more basic. They stand alone. If any of these hypotheses is a true generalization, it would have to be represented by an entry in what I have called the Cosmic Statute Book. If the laws of physics require matter to last for all time, a person who sought knowledge about the past and future duration of matter would expect to find this stated in the Cosmic Statute Book. He would expect the same, if the laws of the Universe had required all matter to come into existence at a specific past time or if they had required it to last until a specific future date. If a building might be completed at *any* future time, the time factor need not be mentioned in the contract document; but when the time to be taken has a limit the date has to be stated.

But (A3) and (B3) are not hypotheses in their own right. They are implicit in the more basic hypothesis that I have called the Principle of Minimum Assumption. [XIII] If (A3) and (B3) are true generalizations, they a not mentioned in the Cosmic Statute Book. If the laws of physics allow an elementary component to originate at any time and to become extinct any time, a clause to say so would be redundant. A person who consulted the Cosmic Statute Book and found no mention of the past and future duration of matter in it would infer that the laws of physics prescribed nothing; he would infer the combination of (A3) with (B3).

In other words, the Symmetrical Impermanence of Matter is a derived hypothesis while all other combinations from the (A) and (B) lists are basic hypotheses. This means that to test the Hypothesis of Symmetrical Impermanence is to test the scope of the Cosmic Statute Book. The question is not what that metaphorical document would have to say about the past and future duration of matter, but whether it would have anything all to say on the subject.

[XIII] *This assertion, obvious though it looks, may have to be abandoned if the theory put forward in Appendix H proves true.*

Chapter 6

The Conservation Laws

The question arises whether the Hypothesis of the Symmetrical Impermanence of Matter violates the great principle of conservation of energy and the other conservation laws. At first sight it certainly does seem as though both continuous origin and continuous extinction do so.

6.1: *Change in the Mass Content of an Isolated Region*
Let us imagine that a region of space can be isolated from its surroundings in such a way that no mass and no energy can pass its boundary in either direction. According to (Al) the total mass and the total energy in this region will then remain constant. According to (A2) they will remain constant now, but would not have done so during that period of the remote past while the material universe was in process of originating. According to (A3) they do not necessarily remain constant at any time; at unpredictable moments new elementary components may originate within the region. If these do not themselves possess inertial mass, it is implicit in the hypothesis that they will eventually synthesize into particles that do possess it. So (A3) implies without doubt that the total mass within the isolated region may increase. Similarly (B3) implies that the total mass may decrease, and as the rates of increase and decrease are assumed to be random the amount of inertial mass in the region at any moment must be indeterminate.

If the region be enlarged until it becomes a fair sample of the whole material universe the argument that its content is not constant in spite of the isolation still holds, of course, but the content is no longer so in- determinate. Like the future activity of a lump of radium, the future content of the region can be predicted, though not from knowledge of external circumstances. It can be predicted only from knowledge of the net rate of random increases.

6.2: *Change in the Energy Content of an Isolated Region*
What is true for mass is also true for energy. But here the situation is more complicated; for the energy that accompanies each individual origin or extinction is a composite affair.

One part of it is represented by the inertial mass itself. For mass and energy are interchangeable, and conversions of the one into the other are today a commonplace of nuclear physics. Hence some of the newly formed mass may sooner or later reappear as the energy of a photon. If, for instance, hydrogen atoms form from the newly originated elemental components, they may in due course combine into atoms of helium, with some loss of mass and release of energy as happens in the sun.

Another part of the newly originating energy is in the electrical fields that surround any charges that are formed. If a negative charge originates in one place and a positive charge in another the configuration formed by the two new charges will contain a certain amount of energy by virtue of the distance between them. The greater this is, the greater the energy in the electrostatic field. So this component of the energy is a variable quantity Its value depends on a fortuitous circumstance, namely, the places in which the two charges happen to originate.

A third part of the energy of each new inertial particle is that possessed by virtue of the position of the particle in a gravitational field. Wherever the particle may originate it must have a store of such potential energy; for it must be within the field of gravitational force of some, though perhaps far distant, accumulation of inertial mass. It can fall on to this, gaining kinetic at the expense of potential energy as it does so, until it can fall no further.

This third part of the total energy must vary enormously between different new particles. Those that originate near a star can fall only a little way and are thus credited at birth with only a little potential energy, while those that originate far out in interstellar space begin their existence with a large store of energy. The kinetic energy with which they eventually arrive at the source of the particular gravitational field in which they have originated will be correspondingly great.

Thus, each origin and each extinction not only adds energy to, or subtracts it from, an isolated region, but the amount of the energy is both an extremely variable and a fortuitous quantity.

What holds for an isolated region of space or an isolated system holds also for the whole material universe. According to both (A3) alone and a combination of (A3) with (B3) the total mass and energy of the universe is continuously increasing. Is this conclusion consistent with the great conservation laws? If it were not, I think that (A3) and (B3) would have to be abandoned. For the conservation laws are too well established, their practical value is too great, for their sacrifice to be contemplated with equanimity. But there is no cause for alarm. All that has to be surrendered are certain misconceptions and certain careless ways of formulating the conservation laws.

6.3: *A Correct Form oj the Principle of Conservation of Energy*
A correct and precise formulation is: *In any system the total energy is not changed by any change in the relation between the component parts of the system.*

It is not, of course, the same for other physical quantities. The total amount of acid in a system may well be changed by a change in the relation between its component parts. Chemical reactions may increase or decrease the amount of acid. Similarly the total potential energy in a system may be changed when the relation between the component parts changes; for some of it may be converted into kinetic energy.

It is in the above formulation that physicists and engineers regard the principle of conservation of energy when they make practical use of it. The would-be inventor of a perpetual motion machine tries to increase the energy obtainable from a system by changing in some ingenious way the relation between its component parts. The principle foredooms his attempt to failure. A scientist knows that changes in the system can influence only the form and distribution of the energy contained in it and not its total quantity, and this knowledge makes it possible for him to work with a balance-sheet and an income and loss account for energy. He does this in effect when he sets up his energy equations.

The principle of conservation of energy tells us that energy income must balance expenditure if capital is to remain constant, but it says nothing about the source of the income or about the recipient of the expenditure. In our experience the income always arrives across the spatial boundary of the system. In a steam engine, for instance, it enters with the steam at the stop-valve. The expenditure is also across the spatial boundary. Some of it leaves through the connecting rod and does useful work in driving whatever machinery is being operated. A larger quantity leaves with the low-grade steam through the engine exhaust. A further part leaves through the hot surface of the engine as heat carried away. All of it was at one moment within space inside the system and is at a later moment within space outside it. The boundary of the system has been crossed.

Such observations have led people to form the following hypothesis: Energy cannot enter or leave a system except by crossing its boundary. But this is not commonly recognized as an hypothesis. It is regarded as a proved and irrefutable fact, which only shows how difficult it is to recognize an hypothesis when one sees one.

6.4: *An Incorrect Form of the Principle of Conservation of Energy*
People have unfortunately combined this unproven hypothesis with the Principle of Conservation of Energy in such a way as to give the principle the following unjustifiable and imprecise form: *The total energy in a system changes only when energy crosses the boundary of the system.* This would be acceptable if one could justify the hypothesis that energy can enter or leave a system only by crossing its boundary, but if energy can enter or leave without doing so the formulation just given is not necessarily true, and is certainly not a correct inference from the Principle of Conservation of Energy.

53

The choice is therefore not between the Principle of Conservation of Energy and the Hypothesis of the Symmetrical Impermanence of Matter. It is between the hypothesis that energy can enter or leave a system only by the specific process of crossing its boundary and that it can do so without crossing the boundary. Both hypotheses are equally compatible with the Principle of Conservation of Energy.

6.5: *A Second Correct Form of the Principle of Conservation of Energy*

Let this principle be given a second, and also correct, formulation: *The total energy in a self-contained system is constant.* For the purpose of this formulation a self-contained system is defined as one into which no energy enters and out of which no energy departs. The second correct formulation says, in effect, the same as the first one. But its interpretation depends on a correct understanding of the concept 'self-contained'.

A correct understanding would, I think, be helped if a distinction were made between a *self-contained* system and an *isolated* one. An isolated system would then be defined as one across the boundary of which energy did not pass. Every self-contained system would, by these definitions, also be an isolated one, but every isolated system would not necessarily be a self-contained one. While the Principle of Conservation of Energy asserts that the total energy in a *self-contained* system is constant, it does not necessarily assert this for an *isolated* system. It would do so only if 'self-contained' and 'isolated' were synonyms. But to claim that they are is to adopt the unproven hypothesis to which I have already referred.

6.6: *An Isolated System is not Necessarily Self-Contained*

By taking the necessary precautions one can ensure that a system prepared in the laboratory can be treated as an isolated one. To do so one must shield the system from external influences and provide its boundary, for instance, with adequate heat insulation. The isolation may not be theoretically perfect, but one makes it as nearly perfect as may be needed for practical purposes. But if the hypothesis of continuous, random, uncontrolled and uncaused origins and extinctions is correct, one cannot ensure that the system will be self-contained, even in theory. For origins and extinctions may occur within it at any moment.

The maximum changes to the total energy content of any system built to the laboratory scale would, however, be too small to be measurable. For practical purposes one can consider any isolated system as though it were also a self-contained one.

This would be far from true on the astronomical scale. There the difference between an isolated and a self-contained system could be very large. Supporters of (A3) interpret the red shift as a measure of this difference.

According to the Hypothesis of Symmetrical Impermanence the whole material universe can be regarded as an isolated system, but it does not meet the definition of a self-contained one. It has, however, been so readily taken for granted that the whole material universe must of necessity be a self-contained system that the notion of continuous origin and extinction has become difficult to accept. It must, nevertheless, be appreciated that it is no more than an unproven hypothesis to assert that the material universe is, by the definition given above, a self-contained system. This is just one of the many sly hypotheses that cause misconceptions. They enter our deliberations so unobtrusively that their entry passes unobserved. Thereupon they clothe themselves in the garments of an irrefutable fact. The disguise is often so excellent that even the most critical do not penetrate it.

Chapter 7

Disappearance Without Trace

7.1: *Introduction*

If, as is postulated by the Hypothesis of Asymmetrical as well as by that of Symmetrical Impermanence, the content of the material universe is always increasing, room must be found for the increase; otherwise the quantity of matter per unit volume would have become infinite. But as, in accordance with the theory of relativity, the extent of the material universe increases with its content, no problem arises here. For a sufficiently large sample of space, the quantity of matter per unit volume must remain constant. In an expanding space matter, one may say, is removed to distant regions as fast as it originates.

But ponderable matter is associated with its manifestations. When it has been removed to some distance some of these can still be observed; what is far away becomes known to us by the traces that it leaves with us. If there were such traces of everything that is and has been, however distant, the number of traces per unit volume here and now would be infinite. It is not infinite and so one must conclude that, when ponderable matter is removed, it disappears without trace.

For objects at astronomical distances only two kinds of trace need to be considered; they are the radiation that the objects send out and the gravitational fields that surround them. Let us consider each of these in turn.

7.2:   *Radiation*

The radiation consists of electromagnetic waves and may or may not be within the visible spectrum; but it is convenient always to speak of it as light. It is caused by events in the microstructure of the objects in which it is generated. A change in the state of an atom, for instance, associated with a change in the orbit of an electron, will generate a pulse of light, a photon, which then travels out into space. It may, on the rarest of occasions, enter a telescope and deliver its message of what has happened to the atom.

Thus light from a distant object brings information just as a telegram does, and for this reason a pulse of light is not inaptly called a signal. But compared with the messages that are carried over telegraph wires, those carried through space by photons can be described only as monotonous. Translated into words the story that they tell would be: 'Now an electron has changed its orbit, now another electron has changed its orbit, now another electron...' and so on, for many millions of years.

Of course we rarely read the signal in that way unless we happen to be astrophysicists. We more often interpret the message as telling us: 'I am a star and I am here'. But the astrophysicists are right and we are wrong. The message is not of what is but of what happens. If there were no changes in the microstructure of the star, we should not have any means of seeing it or of discovering that it was there. As we receive a message to say that something is happening we may safely believe that there is something that it is happening to, but we must not confuse events with things. It is only the events that are observed. The things are inferred.

For much scientific work this philosophical distinction can be ignored but not here. In the most general terms the event is always a conversion of energy. When an electron changes its orbit and sends out a pulse of light, energy is emitted with the pulse. Were it not so, the star would be invisible. We see it only when it converts some of its energy into signals. Physicists have thus come to regard it as axiomatic that what is perceived has been acting prior to its perception as a source of energy and losing energy while it did so. When the object is seen in reflected light, the energy is borrowed. But the visible stars are self-luminous. So long as they signal to us they are using up their capital.

The intensity of the light from a star is almost inversely proportional to the square of the distance of the star and would be quite so if there were no intervening substance to intercept some of it. In fact, less light is received than would accord with the inverse square law, but that does not alter the fact that the amount from each visible star is finite. If a finite amount, however minute, reached us from every one of an infinite number of stars, we should have here and now an infinite luminous intensity. It has to be explained why it is not so.

The explanation is not difficult; it is implicit in the finite velocity of light and the observed expansion of space. As the velocity with which objects move away from one another in an expanding space is proportional to how far they are separated from each other, there is a distance of separation at which the rate of their mutual recession is equal to the velocity of light. At this distance, or at a greater, a light signal emitted from one of them can never reach the other. This limiting distance is called the optical horizon. The radiation received here and now is only from objects that are located within this horizon. Its intensity is finite because the number of stars within the optical horizon and from which signals are received is finite.

(A3), it is to be noted, has sufficient explanatory power to account for the expansion of space and, therewith, for the finite value of the observed luminous intensity here and now. No additional hypothesis need be invented.

57

This can be put in another form. The matter that is removed by the expansion of space can only disappear without trace because the traces are signals of events, and all signals require transmission of energy at a finite velocity.

7.3: *Gravitational Fields*

Let us now consider the other kind of trace, the one left by the gravitational fields that surround all stars. Like the light intensity the gravitational field conforms to the inverse square law. Though this field is very faint when the distance is great, it is finite for a finite distance. The sum of all the traces of gravitational pull from an infinite number of stars would be infinite if they could all act in one place. The curvature of space, which in relativity theory represents a gravitational field, would be such as corresponds to an infinite field intensity.

It is necessary to explain why it is not so. Why is the intensity of the field such as would result from a finite number of sources?

If the gravitational field consisted of events, of signals, each of which acted like a photon and transmitted energy, the answer would be easy. It would be the same as has been found for the finite value of luminous intensity. In that case no additional hypothesis would be needed; one could then infer a gravitational horizon analogous to the optical horizon. Its distance would depend on the velocity with which gravitational pulses travelled. The gravitational effect experienced in any given place would then be only that emanating from stars within the gravitational horizon.

But such an explanation would not be reached by logical reasoning from the traditional hypothesis about gravitation, which is that gravitation is the consequence not of what happens but of what is; that it is an inherent property of every object with inertial mass and persists whether something or nothing happens to the mass. According to this hypothesis gravitation is not (as radiation is) a manifestation of any change at its source. We are told that we should perceive the earth's gravitational pull if nothing whatever were to change either in the earth's macrostructure or in its microstructure. It is known that a luminous flux continues only for so long as it is being renewed and that the renewal requires expenditure of energy, but it is commonly assumed that the gravitational field is kept going without the transfer of any energy whatever. While the amount of energy that the sun is losing by radiation during every second is enormous, the amount that is lost by the sun in keeping the planets to their elliptical orbits is declared to be nil.

Such considerations are bound to lead to the suspicion that the traditional hypothesis about gravitation is untenable and should be replaced by a different one. If the gravitational field is, like the luminous flux, a bundle of impulses, the finite value of the gravitational field is explained, as has just been shown. In that case the sun does lose energy in keeping the planets to

58

their orbits; gravitation is in the same need of continuous renewal as every other detectable manifestation of matter; the axiom holds universally that what is perceived has been acting previously as a source of signals and losing energy while it did so.

It would not conform to the demands of scientific method to put forward this as an alternative hypothesis to the traditional one about gravitation if its sole purpose were to explain one single awkward fact. But it will appear in due course that there is much more to justify the view that the gravitational field does carry away energy and resembles radiation in this respect. However, several further steps of reasoning will have to be taken before this conclusion can be arrived at. So I propose at this moment only to mention one of the many other hypotheses that might perhaps occur rather readily to the mind as a possible means of explaining the finiteness of observed gravitational fields.

Perhaps, it might be thought, the inverse square law requires a correction; perhaps it is only a first approximation to the truth and the real law contains a term that is negligibly small at short distances, but becomes appreciable when the distance is great. If this correcting term represented a repulsion between inertial masses, those that were widely separated would tend to move apart while those that were close together would tend to move towards each other.

Such an hypothesis would provide an easy way out of the dilemma and it might seem plausible at first sight. But, even if such a modification of the inverse square law could be justified (which it cannot be, see Appendix F), it would represent that undesirable thing *mad hoc* hypothesis designed to surmount one particular difficulty. It will appear later that, if Symmetrical Impermanence is accepted, no such desperate measures are necessary. The reason why, in an expanding universe of infinite extent, the gravitational field is constant here and now will emerge, without the need for additional hypotheses, as a logical inference from Symmetrical Impermanence.

PART III

THE ORIGIN AND EVOLUTION OF GALAXIES

Chapter 8

Introduction to Part Three

In the remainder of this book the Principle of Minimum Assumption in general, and the inference from it that I have called the Symmetrical Impermanence of Matter, will be tested by the criterion of explanatory power. It will be shown that one can infer from this principle the answers to a number of questions about our galaxy and those extragalactic bodies that resemble it closely enough also to be called galaxies. It will also be shown in Part Four that one can infer the characteristic properties of gravitation from the same basic principle.

Among the questions to be raised and partly answered are:

Why does the space within our galaxy contain very diffuse hydrogen?
Why does some of the matter in the universe occur in the form of concentrations similar to our galaxy?
Why do the masses of all galaxies lie between finite upper and lower limits?
Why do stars and galaxies rotate ?
Why do many galaxies resemble one another in having the charac- teristic structure for which a dense central core is surrounded by spiral arms all in the same plane ?
Why do the galaxies contain stars ?
(A similar list of the questions about gravitation will be provided in Chapters 21 and 22.)

Not so very long ago the search for answers to such questions would have appeared meaningless. It was thought that stars, nebulae and galaxies just were, and it was taken for granted that they always had been like that. Questions about their origin seemed idle and not worth asking.

A new outlook occurred when it was first appreciated that many processes are irreversible, so that every physical system undergoes certain unidirectional changes. Such a system must have one form or state at its beginning and tend steadily towards another terminal form or state.

With this new understanding of the nature of the physical world questions about the origin and evolution of the galaxies acquired a new meaning. It was realized that these, like all other objects, must be in a process of unidirectional change and so it became natural to want to know as much as possible about the beginning of the process as well as about its development and all subsequent states. In short, we are today no longer satisfied merely to observe the objects that are revealed by our telescopes; we want also to explain them and to discover their past and future histories.

By 'explain' we mean here 'account for in terms of basic physical principles'. When we speak of explaining a phenomenon we mean that the phenomenon can be inferred; that, given certain accepted facts and principles one could predict the phenomenon without needing to observe it. In this sense Dirac predicted the positron, it will be remembered from Chapter 2, and would have been said to have explained it if its discovery had preceded his deductive reasoning. Thus a satisfactory explanation of the particular way in which ponderable matter is distributed must be based on more than an ingenious *ad hoc* hypothesis. It must be based on a sound philosophy, logical inference and a minimum of speculation.

Every one of the nine possible combinations from the (A) and (B) lists that occur at the beginning of Chapter 3 is speculative, though some may regard some of them as more speculative than others. Any such judgements would, however, be merely subjective and depend on the temperament of the person concerned. On an objective assessment one cannot grade the nine combinations into those that are more and those that are less speculative. But the question arises whether, having made one's choice among these nine possibilities, one will need to speculate further. Can one infer from one of the combinations, without the need for any additional hypo- theses, a cosmological model that resembles actuality?

In the rest of this book I shall develop the cosmological model that one infers if one adopts the philosophy expressed by what I have called the Principle of Minimum Assumption, the principle according to which the minimum assumption is always the true generalization about the physicist's world. To meet this principle the model must be based on the combination of (A3) with (B3), on the Hypothesis of the Symmetrical Impermanence of Matter.

It will be found that this model does seem to resemble actuality, where- as every other one, including the model based on the combination of (A3) with (Bl) fails completely.

I have to say 'it seems to', for I cannot exclude the possibility that there may be errors in the reasoning, which will of course have to be checked for these. But if there are errors, they will be of the kind that arises from faulty logic and mathematics, not of the kind that arises from indulgence in unjustifiable speculation.

Let me express the task in a different way. It is to show that all the facts about which the above questions are asked could be discovered with the help of a sound philosophy and accurate reasoning by the inhabitants of a planet permanently enveloped in cloud.

Chapter 9

Interstellar Gas

9.1: *Interstellar Gas must be Falling on to the Stars*
Ponderable matter is very unevenly distributed within our own galaxy. About one-half of it is concentrated in stars. The remainder is in the form of a diffuse gas and occurs throughout the interstellar space. It contains traces of many elements, but consists mostly of hydrogen. As the following considerations will show, its mere existence does not seem to be compatible with either the hypothesis of the past continuous existence of matter or its past finite existence, labelled respectively (Al) and (A2) at the beginning of Chapter 3. But the occurrence of interstellar gas is fully consistent with (A3) and provides some evidence for it.

According to the inverse square law there are gravitational fields in all parts of space, though they are very weak at great distances from any star. Wherever the observed interstellar gas is, it finds itself in such fields and, as it consists of ponderable matter, it must be falling everywhere down the potential gradients of the fields. Every molecule of interstellar gas must be falling with an acceleration that is proportional to the local potential gradient and with a velocity that increases with the time that has elapsed since the falling began.

Let us consider in turn what inferences one must draw about this gas from each of the three hypotheses about the origin of matter: (Al), (A2) and (A3).

9.2: *Objections to (A1)*
According to (A1), unaided by further hypotheses, the gas must have been falling for an infinite time. If it had had an infinite distance through which to fall it would have acquired an infinite velocity. But as the distance between galaxies is finite and the farthest distance through which any molecule can fall must be of the order of half the distance between neighbouring galaxies, every molecule would have completed its fall an infinite time ago. In other words, there would be no more interstellar or extragalactic gas.

Hence (A1) does not suffice to account for the present observation of the interstellar gas without the aid of some additional hypothesis. The falling of this gas on to stars is just one of those irreversible processes that have already been discussed in Chapter 3 and that make it so difficult to reconcile (A1) with observed facts. Here the task is made more difficult by our knowledge of the history of the hydrogen gas after it has completed its fall. When it has reached the star that has been attracting it, it does not necessarily continue to be hydrogen. Some of it contributes to the process known as the synthesis of helium and is converted into that element.

This process occurs when, under the influence of the high pressure and temperature that pertain in a large star, hydrogen nuclei are brought into close association. Some of them then combine to form nuclei of helium and, with still greater pressure and temperature, those of heavier elements. In this process the hydrogen nuclei lose some of their mass, which is con- verted into radiation of a very high frequency. The radiation carries large quantities of energy away with it into outer space.

In order to justify (A1) an hypothesis has to be invented that can explain how the interstellar gas comes to be replenished. One might perhaps explore the notion that some so far unobserved forces are continuously pushing hydrogen out of stars and against gravity. But the potential energy contained in the interstellar gas must be enormous and one would need yet another hypothesis to account for the source of such quantities of energy. Alternatively one might have recourse to the hypothesis of a pulsating universe that has already been mentioned in Chapter 3.

But whatever hypothesis ingenuity can devise to reconcile the inter- stellar gas with infinite past duration, one must remember that more than the mere movement of particles over great distances needs to be accounted for. The existence of interstellar gas would imply, for instance, that the helium synthesis was periodically reversed.

For this to occur the helium would first have to regain the energy that had been radiated away when it was being formed. The protons and neutrons of which the helium nuclei are composed would have to separate and only then could the reconstituted hydrogen leave the star and rise against gravity into outer space, there to distribute itself evenly prior to a renewed fall. To believe that all this can happen is, perhaps, not to believe the impossible; but it puts a severe strain on one's credulity.

9.3: *Objections to (A2)*
As a means of explaining the continued presence of the interstellar gas, (A2) can hardly be called more successful than (Al). True, according to this hypothesis, the gas has not had an infinite time in which to go on falling. But it has been doing so for some thousands of millions of years. During this long period it has been experiencing an acceleration proportional to the gravitational field in which it has found itself. There has been time enough for quite a small acceleration to result in quite a high velocity.

According to (A2), moreover, the universe was a more compact affair at the beginning than it is now, for it is assumed to have been steadily expanding

65

ever since it began. The expansion, it will be remembered, furnishes one of the clocks from which, it has been claimed, the age of the universe can be read off. So the stars must have been closer together in the past than they are now. According to the inverse square law the average potential gradient in which the interstellar gas found itself at the beginning must have been greater and therewith the acceleration that it experienced also greater.

From (A2), unaided by any further hypothesis, one must thus infer that the interstellar gas was falling with a high acceleration to begin with and that, though the acceleration has diminished since then, the velocity has continued to increase. If this were correct, a large proportion of the gas, if not the whole of it, would have completed its fall ere now; and if any were left in interstellar space it would be very unevenly distributed. Regions remote from any star would have been entirely depleted long ago and regions near stars would be surrounded by a comparatively dense cloud of hydrogen of which a minute fraction might be rising in the form of solar flares but most would be falling rapidly. This does not seem to be the distribution of the gas that astronomers observe.

The above is, moreover, not the only difficulty that is presented to (A2) by the uneven distribution of ponderable matter. One must infer from (A2) that some thousands of millions of years ago interstellar space contained a great deal of hydrogen and that much of this has since fallen on to stars. But this hydrogen was accommodated in a smaller universe and must therefore have been at a much greater density than the interstellar gas that we now observe. And as the stars, if smaller, must have been much closer together, their mutual attractions must have been much greater. Why then did they ever separate ? Why did they not fall on to each other? One should expect a universe consisting of many stars in crowded formation with a massive gas distributed between them to shrink and collapse on to itself until it became one single, highly compact, sphere. It does not seem possible to infer from the known laws of physics that the kind of universe that (A2) postulates at the time of the Creation would break up into discrete concentrations, like the observed stars and nebulae. The explanatory power of (A2) is not sufficient to account for the existence of separate concentrations of matter any more than of the more or less even distribution of gas that we observe to exist between these concentrations. I fear that measures devised to save (A2) will have to be as desperate as those devised to save (Al). (A2) will, I think, lose such scientific support as it may still enjoy at the time when this is being written, when scientists begin to ask more questions about the physics of the great explosion that is assumed in (A2) to have occurred once-upon-a-time x-thousand million years ago and said to have begun the expansion of the universe.

9.4: *The Explanatory Power of (A3)*
Here (A3) fares better, both in combination with (Bl) and with (B3). One must

infer from this hypothesis that the interstellar gas is being continuously replenished. Any gas that finds itself in a gravitational field must be falling down the potential gradient, as the water in a brook flows down the valley. But more hydrogen from a region of higher potential must also help to replace what is lost and this must happen throughout space, even in the most distant regions. What is lost to a region by falling out of it is replaced, more or less as the case may be, by new origins. Hence even the remotest region can never become entirely depleted.

If we had no means of detecting the hydrogen that occurs in outer space, we should nevertheless predict from (A3) (and without forming any additional hypothesis) that there would be some. The interstellar gas is, in other words, evidence in support of (A3).

Thus, the inhabitants of a planet surrounded by cloud would, if they adopted (A3) and learnt that we were a part of a galaxy of stars, predict that the space between these stars would contain diffuse hydrogen. They would construct a cosmological model that conformed to actuality in this one respect at least.

They would, of course, also predict that the space between nebulae, extragalactic space, also contained diffuse hydrogen. Observation has so far not succeeded in either confirming or denying this. We can only note the fact that extragalactic hydrogen is one of the inferences to be drawn from (A3). To postulate it is not to adopt an additional hypothesis.

Chapter 10

The Astronomical Landscape

10.1: *Potential Gradients in Extragalactic Space*
It sometimes happens that differences between physical quantities are insignificantly small from some points of view and of decisive importance from others. I propose to show here that it is so for the potential gradients that occur far out in extragalactic space. From the inverse square law of forces one can infer that there is a finite gravitational field everywhere, even in regions very remote from concentrations of matter. But for most purposes the field has a significant intensity only in the vicinity of massive stars; we can usually ignore its existence elsewhere without reaching any false conclusions. But it will nevertheless appear during the next few chapters that small differences in the intensity of the gravitational field in extragalactic space are of great cosmic significance.

In order to appreciate the nature of the very distant gravitational fields let us imagine a space traveller who moves away from the earth, away from the solar system, away from our galaxy, far out into lonely extragalactic space. As he rises from the earth he first experiences the earth's gravitational pull. It is so strong that other fields have a negligible effect. But the earth's field diminishes in accordance with the inverse square law and so a time comes when it is so weak that the field of the more distant sun, hardly noticed before, preponderates. When the space traveller has journeyed still farther and has left our galaxy behind him, the sun's pull becomes as weak as that of millions of other galactic stars. Each of these makes its contribution to the field that remains. It is a very feeble field indeed, but a finite one. If the space traveller were to bring his machine to a halt, he would slowly fall back on to our galaxy.

As the ascent proceeds still farther, this feeble field continues to decrease. The climb is like that of a mountain that is steep at its lower slopes and becomes gentle near the top.

During this imaginary ascent into outer space the space traveller is not only getting farther away from our own galaxy; he is also getting nearer to some other one. He eventually arrives at a region where the faint pull behind him that is exerted by our galaxy equals the forward pull exerted by the galaxy that he is approaching. This region is like that at a mountain ridge; the traveller who reaches it and journeys farther ceases to ascend and begins to descend.

The descent is gentle at first and becomes steeper as the second galaxy is approached.

Another name for intensity of the field is potential gradient; for the field intensity is a measure of the rate of change of potential with distance. This is why one can compare the field intensity anywhere to the gradient on a mountain side. The analogy is not perfect, but it is helpful. Let us therefore begin by considering gravitational potential gradients in the terrestrial landscape.

One can imagine that the earth is surrounded by a number of concentric spherical surfaces of differing radii. Their intersection with mountains will be those lines that appear on topographical maps and are called contour lines. If the shells are equally spaced, each represents a given height above sea level. To a very close approximation each shell is also the locus of all points that have the same potential in respect to the earth's gravitational field. Only a slight difference is occasioned by variations in the value of g over the earth's surface. Let the shells be so spaced that the difference in potential represented by adjacent ones is always the same. For small heights the difference in spacing will then also be nearly constant, just as it is for contour lines. But where the radius of the shells is great compared with that of the earth, equal potential differences will not correspond to equal differences in radii. The potential gradient decreases with the square of the distance from the centre of the earth and so successive shells will be spaced ever more widely as they come to be further out in space. In other words, the vertical journey required to gain a given amount of potential energy increases rather rapidly with increasing distance from the centre of attraction. The diagram in Fig. I illustrates this.

The distance up a sloping mountain side that must be travelled in order to gain a given amount of potential energy is greater the more gentle the slope; and therein lies the analogy. Both in space and up a mountain side a gradient is defined as the distance that an object must be moved in order for its potential to change by a given amount. But in space this distance is measured at right angles to an equipotential surface, while it is measured along the sloping ground when the gradient is on a mountain side.

One calls a landscape flat when movement towards any point of the compass does not result in a change of potential and one calls space flat in the same sense, when movement in any direction, left or right, backward or forward, up or down, does not result in a change of potential. The flatness is in three dimensions here and this makes it impossible to represent the topography of space by a map. In cartography, points of equal potential are connected by lines; in spatial topography they must be represented by curved surfaces. To speak of the spatial landscape is conceptually correct, but it cannot be grasped by the imagination.

If one could do so, one would obtain a picture that differed greatly from any familiar terrestrial landscape.

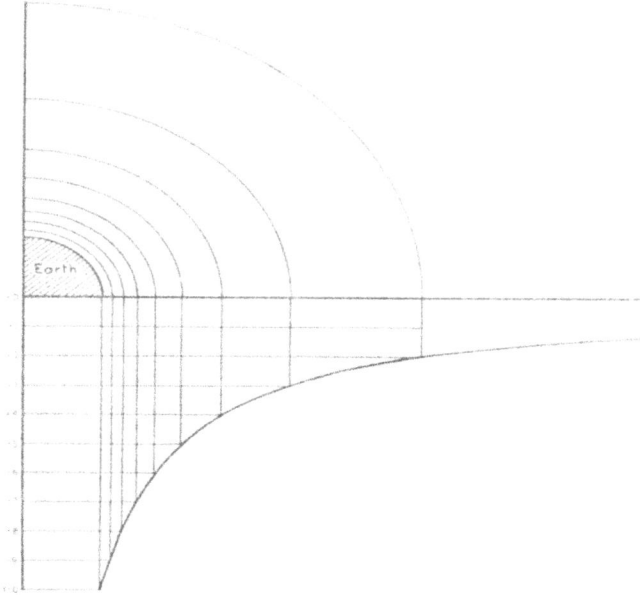

Fig. 1. The Astronomical Landscape near the earth. The diagram shows the radii of successive shells of equal potential difference (one tenth of work required to remove a particle from surface of earth to infinity). The lower curve shows the equivalent gradient in two dimensions.

Let the analogy be pursued to the extent of describing the terrestrial landscape that would be equivalent to the actual spatial one. Regions of space near massive stars would look like very deep, circular wells with only a very slight paraboloid taper towards the bottom. Regions of extragalactic space would, on the other hand, look like a vast table-land so flat that one might be inclined to call it absolutely featureless.

For many purposes it can safely be so regarded. But I propose to show here that the tiny features that occur in these regions should be expected to be significant and, in fact, to determine the origin, even the shape, of galaxies. So instead of thinking of these regions as flat, let us try to thin of them as though they were represented by those relief maps in which vertical heights are greatly magnified.

10.2: *The Structure of Space*
Another name for gravitational potential gradient is the acceleration experienced by a body that is subjected to no other force than that of gravity.

This acceleration is a vector quantity, and so departures from flatness in the astronomical landscape can be described alternatively as differences in the magnitude and direction of the vector that defines the potential gradient or as differences in the value and direction of the accelerations that would be experienced by freely falling molecules of hydrogen.

The vector that defines the potential gradient provides a physical means by which to distinguish one part of space from another. Hence the different accelerations with which, and the different directions in which, particles move in different parts of space are said by relativists to define the physical properties of the space in which the particles find themselves. For this reason it will often be found useful to say that the vectors that represent potential gradients define the structure of space.

These vectors determine the movement of all ponderable matter. It is they that enter into the calculation of the movement of the planets, the earth, the moon. The earth's orbit round the sun is a function of the magnitude and direction of the gravitational potential gradient in which the earth is from moment to moment. It is a gradient that is obtained by combining vectorially the gradients attributable respectively to the sun and those planets that are near enough to the earth to influence its course.

Similar vectors must combine to determine the movement of molecules of hydrogen in extragalactic space and these distant vectors become the proper study for anyone who would understand what happens there. But there is a great difference in the order of magnitude of the gravitational vectors within the solar system and of those in extragalactic space, as there is also in the order of magnitude of the time intervals that are significant.

The potential gradient near the surface of the earth is, for instance, such that a body free to move there adds to its velocity nearly ten metres a second during every second. In a potential gradient of such steepness a short time suffices for a substantial displacement of matter. Distant though the sun is from the earth, the potential gradient of its field at the earth's orbit is still great enough to cause the earth to complete the orbit within a year, which means that the vector defining its linear velocity is reversed in the short time of half a year.

In the comparatively minute potential gradients that occur in extra-galactic space it would take a very long time for a molecule of hydrogen to add ten metres per second to its velocity. But then a very long time is available for the processes that occur there. Those who like to adopt hypothesis (A2) and believe that the whole universe began at a finite moment in the past have recently come to place this moment some thousands of millions of years ago. As has been mentioned already, this period of time has been arrived at from the evidence of several observed irreversible processes, which are assumed to have begun with the beginning of the universe.

71

It will be found here in due course that a time interval of about three and a half million years does have a particular cosmic significance, though not the one attributed to it by supporters of (A2). Even a small fraction of this interval would suffice for molecules of hydrogen subjected to very feeble accelerations to acquire very high velocities and to move over very great distances. Any hydrogen that occurs in extragalactic space must follow these gradients and must, in time, be displaced from one region to another very distant one. It must during the process acquire a large quantity of kinetic energy, a quantity that would cause considerable disturbance whenever the moving mass of hydrogen hit anything.

In short, the structure of space everywhere has cosmic significance. In mathematical terms the important quantity is:
$_0\int^x E$ dx, where E is the potential gradient at a given distance X. The value of this integral can become great in extragalactic space because X can be great there.

10.3: *Astronomical Summits*
Clearly there must be points in extragalactic space at which the gravitational pulls from all surrounding concentrations of mass exactly cancel. These are points of true flatness and also points such that one loses potential when one moves away from them in any direction. They can aptly be called astronomical summits.

An astronomical summit can be defined as a point in space where the potential is a maximum and the potential gradient is zero, in whichever direction it be measured.

Here the analogy to a terrestrial landscape fails. The latter is two-dimensional, while the astronomical landscape is three-dimensional. At a terrestrial mountain top one could increase one's potential by rising up into the air. But on an astronomical summit one cannot increase one's potential by moving in any direction. Just as at the North Pole every signpost points South, so at an astronomical summit every displacement is downwards.

10.4. *Astronomical Reversal Zones*
I propose to give the name 'astronomical reversal zone' to a surface in space at which the potential gradient reverses. At such a surface the acceleration of a particle subjected only to gravitational forces is zero. A projectile could, of course, be shot through the zone. Its velocity would then diminish as it approached the zone and would increase after it had passed through. An astronomical reversal zone is in no sense of the word impenetrable. But gravitation alone can never cause a particle to penetrate it, only to move away from it. To gravitational forces it offers a boundary.

The astronomical summits are points on a reversal zone. What may be called

72

'astronomical passes' are other points on the zone. A pass lies on a straight line between two adjacent galaxies and the position of the pass is given by the inverse square law if the effect of more distant galaxies is neglected. It would be such that

$$(D_1 / D_2)^2 = m_1 / m_2 \qquad \text{......} \qquad (10a)$$

when D_1 and D_2 are, respectively, the distances of the pass from the galaxies and m_1 and m_2 are the masses enclosed by the reversal zones that surround the two galaxies.

It is clearly not necessary for the masses to be concentrated in the galaxies themselves. Equation (10a) holds however the masses may be distributed within their respective reversal zones. The whole mass may be in the galaxy or a part of it may be diffused as extragalactic hydrogen.

While an astronomical summit is a point on the reversal zone where the potential is a maximum, an astronomical pass is a point on the reversal zone where the potential is a minimum, though the potential is less at any point away from the reversal zone on either side of the pass. In this respect terrestrial summits and passes are analogous. Between passes and summits there are ridges and in a terrestrial landscape water will always flow down the side of a ridge and never across it. Similarly, hydrogen will always flow down the side of an astronomical ridge.

The analogy to a terrestrial landscape is, of course, imperfect, as has been noticed already. A mountain ridge ends somewhere in a plain, but an astronomical reversal zone surrounds a galaxy in every direction: left and right, forwards and backwards, up and down. It encloses a three-dimensional volume. In whichever direction one travelled away from a galaxy one could never get round its astronomical reversal zone. If one travelled far enough one would always pass through this zone.

It may be worthwhile to point out for the benefit of the non-physicist that all this is not hypothesis but only what can be inferred from the inverse square law. If one knew the masses of the neighbouring concentrations and their distances from each other, one could calculate where the astronomical summit was, what was the shape of the valleys and ridges that surround it, what the vector of the potential gradient was in every place, how high the pass was over which one could travel from one concentration to its neighbour.

This spatial landscape is not, like terrestrial ones, unchanging. Perhaps one ought to think of it not as a landscape but as the surface of a boiling, viscous liquid. For the topographical features depend on the relative masses and distances of the neighbouring concentrations, and these do not remain constant. In an expanding universe distances are always increasing and a consequence of this must be a smoothing out of the topography. The dome-shaped tops of the astronomical mountains must become flatter as time goes on, at least provided increase of the mass of the neighbouring concentrations does not have a sufficient steepening effect to counteract this tendency. The cosmological significance of this ever-changing extra- galactic landscape will become apparent in due course.

10.5: *The Domain of a Galaxy*
The volume enclosed by an astronomical reversal zone limits the region from which a galaxy draws its supplies and this gives it its cosmological importance. The domain is analogous to a hollow in a terrestrial landscape. The galaxy at its lowest point can then be thought of like a town that nestles at the bottom of the hollow and is surrounded on all sides by a mountain ridge. Only, of course, 'on all sides' has to be taken literally for the astronomical landscape. An astronomical domain has no outlet any- where. Matter within an astronomical reversal zone only disappears by extinction, not by flowing away to somewhere else.

In a terrestrial landscape a hollow is the catchment area for water. Any rain that falls within it finds its way towards the bottom and any rain that falls beyond the enclosing ridge of hills flows away from the hollow and into a different one. Similarly, the volume enclosed by an astronomical reversal zone is the catchment volume for particles of matter that originate within it and for no others.

In the model based on (A3) the gross annual income of a domain is directly proportional to its volume and the volume is, in turn, a direct function of the mass within the domain as the following consideration shows.

Consider two adjacent domains, each containing a galaxy at its centre. Let the masses within the domain be, respectively, m_1 and m_2. However these be distributed, the two masses must act as though they were concentrated at the centres of gravity of their respective domains.

It follows from equation (10a) that $(D_1/D_2) = (m_1/m_2)^{1/2}$
The volumes V_1 and V_2 of the domains must be roughly proportional to the cubes of the distances D_1 and D_2, and so one can write the approximate expression

$$V_1 / V_2 = (m_1 / m_2)^{3/2} \quad \ldots\ldots \quad (10b)$$

This shows that the domain from which a galaxy can attract hydrogen to itself is greater the greater the mass contained within the domain.

10.6: *The Landscape Around an Astronomical Summit*
A mountain summit in a terrestrial landscape is the meeting place of several ridges and between the ridges valleys descend from the summit down to the lower plains. Similarly, an astronomical summit is the meeting place of several astronomical reversal zones. These separate the galaxies by which the summit is surrounded. The galaxies themselves lie at the bottom of astronomical valleys.

Thus the astronomical summit is by no means analogous to a simple, smooth dome. It has a featured structure. From the top ravines extend in all directions, which develop further down into deeper and deeper, as well as steeper and steeper, valleys. Between the ravines there are astronomical shoulders. They descend, like shoulders from a terrestrial mountain, towards the astronomical passes and the gradient rises on the further side of these along the ridges that lead to the next summits.

A moment's thought shows that the number of principal ravines and shoulders that meet at an astronomical summit cannot vary between very wide limits. Every principal ravine points in the direction of one of the nearest galaxies and every principal shoulder points towards an opening between adjacent galaxies. So the number of principal ravines and shoulders is determined by the number of galaxies near enough to have a significant effect on the local structure of space.

Galaxies are scattered about the sky in quite an irregular manner, but if they formed a closely packed regular pattern and were also all equidistant from each other they could occur on the corners of tetrahedrons. There would then be an astronomical summit inside each tetrahedron and this would be equidistant from four galaxies. A straight line from the summit to each of these four would follow the course of an astronomical ravine. A straight line to the centre of each of the four triangular surfaces that bound the tetrahedron would, on the other hand, follow the ridge of an astronomical shoulder.

With such a regular pattern there would therefore be four ravines and four shoulders. Gentle ditches along the latter would lie on lines pointing to more distant galaxies, but the nearest of these would be well over twice the distance from one of the corners of the tetrahedron. It will be seen later from equation (12f) in Section 12.4 that the potential gradient varies inversely as the cube of the distance, so the effect of all such more distant galaxies is negligible.

The regular tetrahedron would provide the most compact spacing. It cannot be typical of the actual irregular pattern; and so an astronomical summit must normally be surrounded by more than four neighbouring galaxies. One might expect a more frequent pattern to approximate roughly to a cubic arrangement. From a summit inside a cube eight galaxies are equidistant and so there would be eight ravines, while six shoulders would point towards the six sides of the cube. A spacing that led to a greater number of ravines and shoulders could occur, but would not be very common.

Chapter 11

The Pressure of Radiation

At this stage of the present study the question may perhaps be raised whether the pressure of radiation can significantly influence the movement of ponderable matter in extragalactic space. Apart from gravitational forces it is the only force known to act there over great distances. Its effect is large enough to move metal discs in suitably constructed laboratory apparatus. It serves as the most precise means of measuring the power transmitted by a wave guide. It has recently become a major field of study by physicists and engineers.

I cannot believe that any of those who have devoted their serious attention to the pressure of radiation will regard it as having any appreciable influence on the movement of any extragalactic particles, great or small, or indeed on any interstellar particles within our own galaxy. So I should hesitate to introduce the subject at all were it not that some, who are non-specialists in this field, do seem to entertain such a notion.

The view has, for instance, received the backing of considerable scientific authority that the pressure of radiation may have carried germs of living substance away from a planet belonging to some distant constellation and deposited them on our earth. This notion has been urged to account for the way terrestrial life began. On asking an astrophysicist of repute, again, how the interstellar hydrogen of our galaxy can be accounted for, he told me that, according to one view held seriously by himself and others, this gas had been projected into outer space by the pressure of radiation that emanates from bright stars.

So there seem to be some who believe that the pressure of radiation can transport small particles of matter *from* outer space towards our bright solar system and others who believe that the same pressure can transport small particles *into* outer space away from a bright solar system. The two views are hard to reconcile with each other. But the effort to do so can be spared. For I shall show below that, when considering the movement of small particles in outer space, the pressure of radiation can be disregarded. It would, however, seem that current opinion about it cannot. Hence the subject must claim our attention for a short while.

Let the question take this form: When, at any place in extragalactic space, the vector that represents pressure of radiation from a given star. nebula or galaxy is added vectorially to the vector that represents the gravitational field of the same star, nebula or galaxy what is the direction and magnitude of the resulting vector?

The answer provides no difficulty. Both forces operate along the same line, the one being a push away from the source and the other a pull towards it. The resultant vector has therefore the same direction as the two of which it is the sum and points in the direction of the preponderating force. The effect of pressure of radiation is, in other words, to reduce locally the effect of the gravitational constant, G, by a certain amount but not to change the *direction* of the field. But should the pressure of radiation preponderate anywhere, it would change the sign of the field. If it did that, things would fall off a bright and massive body instead of on to it.

Like gravity, pressure of radiation follows the inverse square law. Both forces are therefore greatest in the vicinity of the body that constitutes their source and both diminish in the same ratio with increasing distance. If the one force preponderates over the other near the source, it does so everywhere else. If the pressure of radiation is not enough to overcome gravitation at the surface of a star, it is not enough to overcome it anywhere.

The gravitational pull of a star is proportional to its mass and the radiation push is proportional to its brightness. Those who believe that pressure of radiation has a significant influence on the movement of interstellar particles must therefore assume that the ratio of brightness to mass of at least some stars suffices to cancel G. Does it?

We know that the pressure of radiation at the sun's surface is not sufficient to overcome gravitation there. If it were, hydrogen would be continuously rising against gravity from the whole of the sun's surface. The sun would not hold together. In fact, any star bright enough to throw its substance into outer space would gradually disappear like a drop of water that evaporates into water vapour. And the sun is a star of average brightness. So we may safely conclude that the pressure of radiation does not significantly affect either the position or the steepness of any of the astronomical features that have been discussed above. The astronomical summits, the passes, the boundaries must be almost as they would be in a world devoid of radiation.

We also have to reject the notion that the observed interstellar gas has moved into outer space against the force of gravitation. It is but one of those notions that seem attractive only so long as they are not subjected to the test of quantitative thinking. The notion that the much heavier germs of living substance have been moved by faint starlight not only against the gravitational field of the planet on which they originated during the early part

78

of their journey but also against the sun's fierce brightness during the latter part is, of course, quite untenable. Indeed I find it hard to feel otherwise than flippant about those who tell us that, in all probability, the pressure of radiation brought the germs of living substance to our earth. Do they claim, perhaps, that it happened at night ? Can they have forgotten that the sun is still shining when it is night time on the part of the earth where they happen to be ?

Chapter 12

How Might a Galaxy Begin?

12.1: *Qualitative and Quantitative Considerations*

The significance of the astronomical landscape that has been described in Chapter 10 cannot be appreciated without consideration both of its qualitative and its quantitative aspects, i.e., without its discussion in both words and mathematical symbols. The words are needed to express meaning and the symbols to express magnitude.

It happens to all of us occasionally that we reach a wrong conclusion through neglect of the one or the other of these aspects. Sometimes our mathematical reasoning is quite sound and we err by wrongly interpreting the symbols, while at other times we are quite clear as to what we mean but we fail to notice that our purely qualitative reasoning leads to a result of the wrong order of magnitude. Only mathematical analysis can correct such an error.

For these reasons an investigation is often best conducted in two stages, of which the first is qualitative and the second quantitative.

The non-mathematical investigator cannot follow beyond the first stage, and this may on occasion leave him quite unenlightened; for there are recondite fields of study where the mathematical expressions cannot be translated into ordinary language. But it is hardly so here. I have planned to limit this inquiry almost entirely to the first, qualitative, stage and shall introduce only that bare minimum of mathematics that is needed to give precision to the reasoning. This minimum is unavoidable when one seeks to discover whether the quantities involved, the forces, the distances, the velocities, are of the order of magnitude from which one would infer a cosmological model that resembles actuality.

To help the non-mathematical investigator of the subject I shall precede the mathematical treatment by a short qualitative discussion.

12.2: *The Critical Potential Gradient*

The expression 'astronomical landscape' is no more than a metaphorical way of saying that the acceleration with which things fall in outer space, and the direction in which they fall, change from place to place. Using the same metaphor, physicists speak of a potential gradient and say that this may vary in magnitude and direction. Therewith they suggest a slope down which objects may fall.

To speak thus is to present a two-dimensional analogy to a three-dimensional reality and is in this sense imprecise. But the three-dimensional reality is difficult to visualize and this difficulty may lead the investigator into error. On the other hand, I do not think that it can cause any serious misapprehension if I use the language here of the two-dimensional analogy, so I propose to do this.

Molecules of hydrogen in extragalactic space are always on a metaphorical slope, such as corresponds to this analogy, unless they happen to be precisely on the top of an astronomical summit. These molecules are therefore always falling in one direction or another. Where galaxies on opposite sides of a molecule are attracting it in opposite directions it is moved only by the difference between these opposed forces, forces that are themselves very weak. Hence the acceleration that the molecule experiences is small; but it is finite, except at the mathematical point called the summit.

While molecules are continuously falling down the slopes around a summit, no molecules are falling on to this region, for there is no greater height from which they can fall. For a cosmological model based on (A1) or (A2) the mountain would be entirely depleted of matter.

Not necessarily so for a model based on (A3). Although hydrogen cannot fall on to an astronomical summit from anywhere else it does *originate* there as everywhere else. It depends on the gradient whether the gas near the summit is becoming more dense or more tenuous. If the rate at which particles fall down the slope is greater than the rate at which new matter originates there, the region will be almost entirely depleted of gas; such as is left there at any moment will consist entirely of particles that have recently originated and are in process of falling out of the region. But there must be one particular potential gradient at which the rate at which particles fall away equals the rate of new origins. This can be called the critical gradient.

In any region where the gradient is below the critical value, a cloud will form of which the density will be a function of the gradient. Such a region must occur around an astronomical summit provided this is flat enough.

In an expanding universe astronomical summits must become flatter and flatter as the neighbouring galaxies recede further and further from them. It can therefore only be a matter of time for the critical gradient around an astronomical summit to be reached. When, in other words, neighbouring galaxies have been removed to a certain distance, which can be called the critical one, cloud formation must begin on the astronomical summit that lies between them. As the distance between the neighbouring galaxies further increases and the gradient around the summit becomes still flatter, the cloud must then become increasingly more dense.

The critical distance depends, of course, on the masses of the neighbouring galaxies, but it will eventually be reached for all of them provided only that the masses of the retreating galaxies do not increase at a rate sufficient to prevent this from happening.

12.3: *Significant Changes in the Astronomical Landscape*
It has been pointed out in Chapter 10 that the astronomical landscape is not stationary. The contour lines change with changes in the position of gravitational masses. The significance of this is great in its effect on the cosmological model that is to be inferred from Symmetrical Impermanence. One must, for instance, take into consideration the gravitational field of the newly-formed cloud. This becomes an increasing centre of force as it grows and is more and more effective in attracting hydrogen from outer space. A particle that was falling away from the cloud when this began to form will, at some time, reverse its direction and fall on to the cloud. This then grows by attracting matter from outside as well as by new origins within the region that it occupies.

To say that a particle now falls towards the cloud that previously fell away from it is, in metaphorical language, to say that the gravitational slope has been reversed. A space traveller who was in the cloud would have to climb against a potential gradient to get away from it. In other words, the newly formed cloud acquires an astronomical reversal zone of its own.

So long as the cloud is small and very tenuous the distance from the centre of the cloud to its astronomical reversal zone, and therewith the domain of the cloud, must also be small. One may almost picture the astronomical summit with a cloud on it as having the shape of a volcano, i.e. as a mountain top with a slight crater-like depression. The shape of this must conform to that of the cloud and grow with it so that, while it is the local topography that first moulds the cloud, the cloud eventually moulds the topography. It does so with increasing vigour, causing the depression to become ever deeper and steeper and thereby exerting an increasing attraction on the gas that surrounds it.

The difference between the landscape before the cloud begins and after the cloud has grown has some resemblance to the difference between a photographic negative and positive. The effect of the growing cloud is that peaks may become craters and eventually develop into deep wells, shoulders become valleys, valleys become ridges. As the reversal of the potential gradient, the astronomical reversal zone, travels out into distant space, what began as the lip of a crater comes to be more analogous to an extensive ridge. As I have said before, the astronomical landscape is a changeful one.

Should, in the course of time, the cloud acquire a mass equal to that of the older and more distant galaxies, this ridge will be half-way between, two concentrations: the cloud and the older galaxy. The point of lowest potential along this is on a straight line connecting neighbouring galaxies and is analogous to a mountain pass, as has been explained in Chapter 10.

Particles will fall away from this either in one direction or the other, according to the side of the pass on which they occur. From the top of the pass a space traveller could either descend to one of the concentrations or move at right angles to the direction of descent along the ridge, where he would be ascending a gentle slope until he reached a new astronomical summit.

12.4: *Mathematical Introduction to Cloud Formation*
The reason why a small amount of mathematical treatment cannot be omitted from the reasoning by which the cosmological model is inferred that is implicit in Symmetrical Impermanence is that the order of magnitude of the quantities inferred cannot be assessed by any other means. Hence, a brief discussion in mathematical terms is given below of the conditions that make cloud formation round an astronomical summit possible. It will emerge from quantitative considerations that the model to be inferred from the combination of (A3) with (B3), the Hypothesis of Symmetrical Impermanence, leads to a model that resembles actuality, not only in broad outline, but also in matters of detail.

Let distance be regarded as positive when measured in a direction away from an astronomical summit. The acceleration A of a particle is then also positive in that direction. The potential decreases with increasing distance from the summit, so the potential gradient E around the summit is negative. Numerically the potential gradient and the acceleration of a particle at a given point are equal, so one can write $A = -E$.

There are particular conditions for which the rate at which particles fall away, down the gradient around an astronomical summit, equals the net rate at which particles originate within the region around the summit, the net rate of origins being the excess rate of origins over that of extinctions in the region. I shall call these 'critical conditions' and define them by the suffix c. Thus $A_c = -E_c$ is the critical average acceleration experienced at the critical gradient E_c. The critical distance from the summit at which this occurs is r_c and the critical average velocity with which particles reach the surface of a sphere with radius r_c and centre at the astronomical summit is v_c.

Let n be the net rate of origins within the sphere per unit volume and time and let N be the density of gas at the surface of the sphere expressed as number of particles present there per unit volume. The number of particles that originate within the sphere in unit time is $(4/3)\pi n r_c^3$ and the number that cross the boundary in unit time is $4\pi N r_c^2 v_c$.

To assume that the summit is a smooth dome, lacking topographical features, would be incorrect. But no mistake is made if one considers conditions for one particular radial direction. For that direction the con- ditions are the same as if the summit were dome shaped, as implied by the above expressions.

The critical conditions are defined as those for which the content of the sphere remains constant, which gives:

$$(4/3) \pi N r_c^3 = 4\pi N r_c^2 V_c$$

from which

$$V_c / r_c = n / 3N \qquad \text{......} \qquad (12a)$$

An expression is needed for V_c in terms of r_c and the shape of the slope. This shape is defined by saying that the acceleration at a given distance r from the summit is $A = -E = \varphi(r)$.

Consider a particle that originates at distance r from the summit. Let its velocity be v_s, when it reaches the surface of the sphere. By definition

$$A = dv/dt = dv/dr \cdot dr/dt$$
$$Adr = vdv \text{ and}$$
$$\int_r^{r_c} \varphi r \, dr = \int_0^{v_s} v \, dv = 1/2 \, v_s^2$$

If the astronomical summit has the shape of a parabola, one can put $\varphi(r)=kr$ and the above equation becomes

$$\int_r^{r_c} k r \, dr = 1/2k(r_c^2 - r^2) = 1/2 \, v_s^2 \qquad \text{......} \qquad (12b)$$

This expression holds for any particle that originates at distance r from the summit. To obtain the average value of v_s which is v_c, one must obtain the average value of r.

This is obtained by considering a shell with its centre at the summit and radius r, smaller than r_c. Let its thickness be dr. The volume occupied by this shell is $4\pi r^2 dr$ and the net number of particles that originate within the thickness of the shell is $4\pi n r^2 dr$. Each particle reaches the surface of the sphere after falling through the distance $r_c - r$. The total number of particles that fall in unit time multiplied by the distance through which each falls is

$$4\pi n \int_0^{r_c} r^2(r_c - r) \, dr = (1/3) n\pi \, r_c^4$$

The total number of particles originating within the sphere in unit time is $(4/3)\pi\, r_c{}^3$, so the average distance fallen by a particle is

$(r_c - r_m) = \{(1/3)\pi\, n\, r_c{}^4\} / \{(4/3)\pi\, n\, r_c{}^3\} = r_c / 4$

Where r_m is the mean value. From which:

$r_m = (3/4)r_c$

In equation (12b), $r = r_m$ when $v_s = v_c$. Inserting these values one obtains

$v_c = \sqrt{(7k)}\, (r_c / 4)$ ......   (12c)

When this value is inserted in equation (12a) one obtains

$n/N = (3/2)\sqrt{(7k)}$

or
$N = 2\pi / \sqrt{(63k)}$ ......   (12d)

This shows that for a parabolic slope the density, N, depends only on the value of k and is independent of the radius. A parabolic slope is therefore a region in which the gas density is uniform and increases as the slope flattens and k decreases.

It will be shown below that the actual slope around an astronomical summit is nearly, but not quite, parabolic. The calculation is somewhat complicated for the general case, but it will suffice for the present purpose to consider a gradient that lies on a straight line between two concentrations having equal masses, m; to consider, in other words, a pass instead of a summit. The error made by doing so is probably not negligible by any means. But the simplification of pretending that a summit is just like a pass facilitates understanding of the broad outline of the problem. The gradient at a pass is

$E = Gm[1 / (D + r)^2 - 1 / (D - r)^2]$

Where D is the distance from the astronomical summit to one of the centres of mass. The above formula can be written more simply

$E = -4GMDr / (D^2 - r^2)^2$ ......   (12e)

When r is small compared with D this becomes nearly

$E = - GMr. / D^3$ ......   (12f)

Putting $4\,G\,m\,/\,D^3 = k$ gives

$A = -E = kr$

which is the expression for the parabola defined above.

This means that, when k has dropped to the value at which a cloud can begin to form at an astronomical summit it has nearly dropped to the same critical value at a little distance from the summit. As space expands and k decreases further, it soon will acquire the critical value there. Hence the cloud around an astronomical summit is of nearly, but not quite, uniform density. It is a little denser at the centre than at its fringe; and it grows outwards rather rapidly. The difference between equation (12e) and (12f) is a measure of the rate of spread. This is illustrated by the diagram in Fig. 2.

It should be noted that k is inversely proportional to the cube of the distance from a galaxy, which means that, in expanding space, the flattening of the parabola around an astronomical summit is rather rapid.

This theme should not be left without a hint as to how to arrive at the equation that must replace (12e) when it is necessary to express accurately the gradient around an astronomical summit. Equation (12e) has been obtained by superimposing the potential gradients of two equal masses The true equation must represent the superposition of all masses neai enough to have an appreciable effect. Their distribution is in three dimen- sions, which makes the calculation complicated. But none of them is as near to a summit as to the nearest pass. As those on opposite sides of a pass have potential gradients of opposite sign and so largely cancel eaci other, so do masses on opposite sides of a summit. Hence the gradient

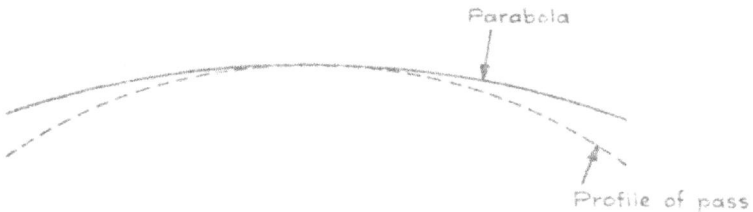

Fig. 2. Shape of an astronomical pass

around a summit is flatter than that in a corresponding position near a pass. One may expect this to be expressible by the addition of a term dependent on r/D. This would be done if equation (12e) took some such form as

$E = - \{ G\,m\,D\,r\,/\,(D^2 - r^2)^2 \}\,\varphi\,(\,r/D)$ ...... (12g)

Equation (12g) can hardly be of the correct form, but it serves to suggest the way in which a number of masses in three dimensional arrangement. might be expected to influence the potential gradient. As the factor $\varphi \, (r \, / \, D)$ must reduce the value of E for it to mark the change from a pass to a summit and as r is much less than D, the function must be a positive one. In other words the function must vary directly with $r \, / \, D$.

The fringe of the cloud has been mentioned above and needs to be precisely defined. This choice of term should not mislead anyone into picturing a sharp outline between a region of high and one of low mass densities. Our imaginary space traveller would not observe any frontier on passing through the fringe. Nevertheless it can be defined quite precisely in words. The fringe of a cloud is a shell of indefinite thinness that surrounds the centre of the cloud and is so placed that the rate at which matter falls down the potential gradient balances the rate of origins. On one side of the fringe the rate of falling exceeds the rate of origins; the region becomes increasingly depleted. On the other, where the gradient is flatter, the rate of origins exceeds the rate of falling away. If N is the number of particles per unit volume, the fringe is the shell for which $dN \, / \, dt = 0$.

Chapter 13

Growth of an Extragalactic Cloud

13.1: *The Problem Defined*
In the last chapter it was shown that, if matter originates continuously, as is asserted in (A3), a cloud must occasionally form in extragalactic space. But this in itself does not go far towards proving that a cosmological model based on (A3) resembles actuality. To prove this one must do much more: One must show that in such a model the undifferentiated accumulation of hydrogen that I have called a 'cloud' will evolve into the kind of structure to which the name galaxy has been given. The simple accumulation of hydrogen must acquire a mass of the same order of magnitude as is possessed by the observed galaxies; its distance from its nearest neighbours must be of the same order of magnitude; it must behave in a similar manner; it must have a similar volume, density, shape, detailed structure. Why, one is led to ask, should all this happen?

13.2: *The Variables that Influence Growth*
There is a substantial risk of drawing wrong inferences when constructing a cosmological model on the basis of any selected hypotheses, and one reason for this is that a large variety of circumstances must influence the evolution of an extragalactic cloud, and that most of them interact with each other in complicated ways. Hence great care must be taken to con- sider all these circumstances and to assess their effects. It will help towards the understanding of what is to follow if they are listed here. Let it be pointed out, therefore, that what happens to a cloud after it has begun to form on an astronomical summit must depend on:
(a)   The rate of origin of new matter in the region, which is at present unknown.
(b)   The rate of extinction of matter in the region, whichis also unknown.
(c)   The known expansion of space, which causes the gravitational pull of distant nebulae on the substance of the cloud to decrease with time.
(d)   Changes, if any, in the masses of neighbouring distant galaxies, which must occasion changes in the gravitational pull that they exert on the substances of the cloud.
(e)   Changes in the density of the cloud.
(f)   The gravitational field of the cloud itself, which must fundamentally influence the distribution of forces in the region where the cloud is, reversing the local potential gradient and causing particles to fall towards the cloud that previously fell towards a distant galaxy.

(g) Irregularities in the potential gradient around an astronomical summit.
(h) Shrinking of the cloud under the influence of its own gravitational field.
(i) Rotation of the cloud, which introduces centrifugal forces.

The length of the above list serves as a warning against premature treatment of the subject in mathematical terms. It is at the stage of qualitative consideration that one is best able to ensure that none of the interacting circumstances is overlooked, and so I shall here, as in Chapter 12, restrict mathematical treatment to the unavoidable minimum. The cosmological model can be checked for its detailed quantitative resemblance to actuality more easily after, than before, its broad qualitative features have been inferred. The small amount of mathematics introduced here will therefore be no more than is needed in order to test whether certain inferred quantities are of the right order of magnitude.

13.3: *The Cloud's Income and Loss Account*
If the inferred cosmological model is to resemble actuality, one thing that must certainly happen to it is that the incipient cloud must grow in extent and density during its transition from cloud to galaxy, and for this to happen it must receive more particles than it loses. Let us consider both the cloud's sources of income and its sources of loss. They can be thought of as analogous to profit and loss in accountancy; but here income and loss are more precise terms. The sources and amounts of both differ greatly according to whether the Hypothesis of Asymmetrical or of Symmetrical Impermanence is used as a basis.

Sources of Income: According to both hypotheses there are two sources of income. The first is origins within the region occupied by the cloud, and the second is existing matter that the cloud attracts from its surroundings by virtue of its own gravitational field. According to both hypotheses the first source of income is constant per unit volume. (This is probably obvious, but the reason will be fully given in Chapter 15.) But the amount of income from origins is not the same in the two hypotheses. According to Asymmetrical Impermanence it is always the 500 atoms of hydrogen per cubic kilometre per year estimated by McCrea. But according to Symmetrical Impermanence this number is the difference between the rate of origin and the average rate of extinctions per unit volume for the whole universe. The gross rate of origins is greater and the net income depends on the local rate of extinctions, as will be explained in Chapter 14.

Remembering the two distinct sources of income, one can speak of growth by the origin of new matter and growth by the capture of existing matter. While the cloud is still small and tenuous it has little mass and cannot attract matter towards itself in competition with the more massive neighbouring galaxies; so growth by capture cannot occur. On the contrary, the cloud does not even retain all the particles that originate within it. Some of these are attracted away towards the neighbouring galaxies; in the metaphor of an astronomical landscape they fall away down the slope. But in so far as origins cause the cloud to become more massive, it competes more and more successfully with the distant galaxies and growth by capture occurs eventually and then at an increasing rate. Growth of an extragalactic cloud thus occurs in two successive stages. During the first it depends entirely on the excess of origins over extinctions in the region occupied by the cloud. At the second it depends to an increasing extent on capture of particles from beyond its fringe.

Sources of Loss: Two sources of loss are postulated by Symmetrical Impermanence: loss by falling down the slope and loss by extinction. But only one source of loss is postulated by the Hypothesis of Asymmetrical Impermanence, i.e. the gravitational fields of the neighbouring galaxies, which are the cause of the potential gradient down which some of the particles in the cloud fall during the first stage of growth. When the second stage has begun, this source of loss ceases and there are the two sources of income: the 500 atoms of hydrogen per cubic kilometre per year and the particles captured from outside the cloud.

13.4: *Income and Loss During the First Stage of Growth*
Now 500 atoms of hydrogen is not much. It would not go far towards fattening a microbe. Even if the 500 were all retained in the cloud from the moment when it began to form, it would take about four million million years before the cloud gained one molecule of hydrogen per cubic centimetre, and this number would still leave the cloud very tenuous.

How minute a quantity 500 atoms of hydrogen is can be appreciated from another consideration. Some advocates of hypothesis (A2) have been claiming recently that the universe was created about seven thousand million years ago. A cloud that was acquiring 500 atoms of hydrogen per cubic kilometre per year during this span of time would now have a density such that a volume equal to that of the earth would weigh about one gramme.

But in our model the rate at which the cloud gains mass is much less than even this very low value. For only a small fraction of the newly originating atoms are retained in it. When the cloud is just beginning to form, nearly the whole lot of them are lost, for the potential gradient is such that the number of particles falling down the slope is just, but only just, less than the number that originate in the region. In this model the periods of time taken by a volume equal to that of the earth to increase its mass by one gramme is much more than seven thousand million years. It is not until the gradient has become nearly flat that most of the atoms are retained and that this maximum rate of growth is achieved.

In this model the rate of flattening is, moreover, slow, for the potential gradient is subjected to two opposing tendencies. The first is the expansion of space. As the neighbouring nebulae, which are the cause of the potential gradient, move further away, the force per unit mass with which they attract particles out of the region diminishes, which is another way of saying that the metaphorical slope flattens.

But for the model based on Asymmetrical Impermanence there is an opposing tendency. It will shortly be shown that in this model the retreating galaxies are themselves rapidly becoming more massive. Hence the flattening effect of increasing distance is at least partly offset by the steepening effect of the growing masses of the galaxies. It does not seem certain that the potential gradient decreases at all in this model; but if it does it must happen very slowly. However, it will be shown in Appendix B that this opposing tendency need not operate for the model based on Symmetrical Impermanence.

There is yet another circumstance that retards the growth of the cloud during the first stage. This is the fact that the loss rate is a direct function of the density, as can be inferred from equation (12a) in Section 12.4. For every gradient there is a limiting density above which the rate of loss down the slope exceeds the rate of income. The cloud cannot become more dense until the gradient has become flatter.

The cumulative effect of these various retarding factors is that, in our model the first stage of growth is exceedingly slow. The cloud will, however, become significantly more massive during the second stage of growth, when it captures matter from outside. This will cause its density to increase: and the density will also increase as the cloud shrinks under its own gravitational field. One must expect the cloud to be much more extensive at first than the galaxy into which it evolves.

For the model based on Symmetrical Impermanence the rate of income during the first stage of growth is greater by an amount at present unknown. The 500 atoms of hydrogen per cubic kilometre per year (or whatever the correct figure may be) are a net and not a gross rate. They are the average excess of origins over extinctions for the whole of the observable universe and are not relevant to any particular locality.

The region where the cloud has just begun to form has been depleted of matter until that moment, and so there cannot be many particles to become extinct in the region and the local net rate of origins must be very nearly equal to the gross rate. Hence the model based on Symmetrical Impermanence permits a more rapid growth during the first stage. One could only assess the significance of this distinction if one knew the gross rates of origins and extinctions.

13.5: *End of First Stage*
The moment when the first stage of growth is ended and the second stage

begins can be defined precisely in terms of the fringe of the cloud. This is the surface around the incipient cloud at which the income from new origins balances the loss from extinctions and from falling away down the slope. (See page 102.) It is the boundary within which the density of the cloud increases and it is associated everywhere with a specific potential gradient, which I have called the critical one. The end of the first stage is the moment when particles just beyond the fringe of the cloud reverse their direction and fall towards, instead of away from, the cloud. This happens at the moment when the potential gradient at the fringe reverses, when the gravitational pull exerted by the cloud at its fringe just suffices to cancel the pull exerted by a neighbouring galaxy. It is the moment when the astronomical reversal zone of the cloud coincides with its fringe. After the cloud has become massive enough for this to happen, no more particles fall out of it.

In the metaphor of an astronomical landscape, the zone at which the potential gradient reverses is, it will be remembered, a ridge from which the ground slopes to either side. This ridge begins as the lip of a crater at the astronomical summit and moves gradually outwards as the mass of the cloud increases. The diagrams in Fig. 3 will help to make the course of events clear.

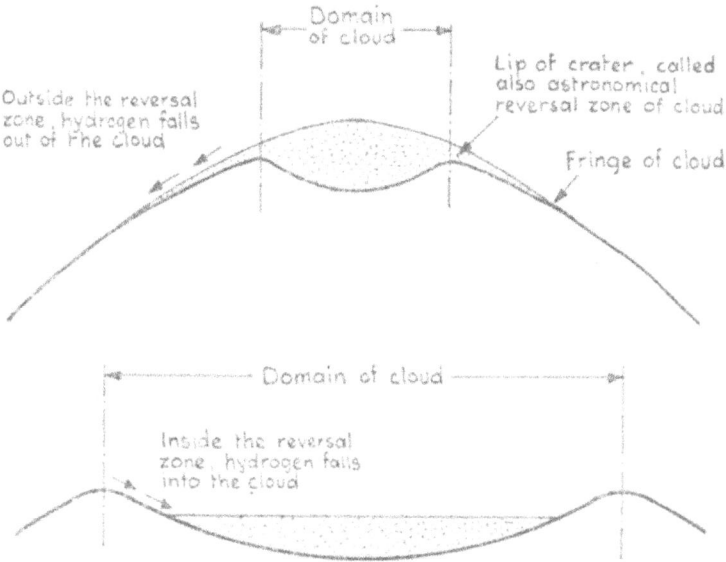

Fig. 3. The two stages of growth of an incipient galaxy. *Above:* First stage *Below:* Second stage.

At first, when the cloud has only just begun, the crater is small. There is cloud on each side of its lip, as is shown in the diagram that represents the first stage. Particles do not fall out of this part of the cloud that is within the crater;

but they do fall away from the outer part. But as the lip moves outwards, becoming more like a mountain ridge, less and less of the cloud is outside it and the loss rate from falling down the slope decreases. The moment when the ridge, or reversal zone, reaches the fringe of the cloud is the moment when loss from this source comes to an end and the second stage of growth begins. As the cloud's mass increases still further, this zone moves out beyond the fringe. The situation is then as shown in the diagram for the second stage. There is a growing region surrounding the cloud from which particles fall towards it, and its mass increases by capture and not only by origins within it.

Should the cloud eventually acquire a mass equal to that of a neighbouring galaxy, the boundary will be half way between the cloud and this galaxy; it will, as mentioned already, be analogous to a mountain pass over which one must travel in order to go from the cloud to the galaxy. A model that resembles reality must be one in which the reversal zone will eventually arrive at approximately such a half-way position, and this depends in turn on the rate at which the slope around the summit flattens. It will be shown later, when the subject is treated quantitatively, that by astronomical standards this rate is very rapid.

In this model the incipient cloud remains tenuous and spreads quickly until the product of its density and its volume suffices to attract particles that are just outside it and that would, before that moment, have been attracted towards the neighbouring nebula. Thereafter its volume may decrease by shrinking, but its mass per unit volume must increase at a greater rate.

Chapter 14

The Average and the Equilibrium Density of the Universe

The average density within an astronomical domain is the total mass within the domain divided by the volume of the domain. It will be found to be an important quantity. To appreciate its significance one must relate it to two other densities. These are the average density of the whole universe and a quantity that I propose to call the equilibrium density.

The meaning of the former is easy to understand. If one considers a region of space large enough to be a fair sample of the whole material universe, the average density is the total mass in this region divided by its volume. To be a fair sample the region may have to be substantially larger than the domain of any particular galaxy.

The equilibrium density is a different quantity and depends on the relative rates at which matter originates and becomes extinct. To make this clear let me recall the basic features of hypotheses (A3) and (B3).

In (A3) no assumption is made to associate the origin of a particle with anything in the existing state of affairs. There is therefore no reason why a particle should originate in one specific place rather than in any other, and the minimum assumption is that every region of space has at any given moment an equal probability of being the birthplace of a particle.

Similarly no assumption is made to associate the extinction of a particle with anything in the existing state of affairs, and so there is no reason why a given particle should become extinct at any particular moment. The minimum assumption is that every particle has at that moment an equal probability of becoming extinct. If one considers any lump of matter, a certain constant fraction of it becomes extinct every year. What- ever the lump may consist of, wherever it may be, to whatever influences it may be subjected, it will lose half of its mass by extinction during a finite, though at present unknown, number of years. On the analogy of the expression 'half-life of a radioactive substance' this must be called *the half-life of mass.*

From this essential aspect of Symmetrical Impermanence one can thus infer that the rate of origins is constant per unit volume and the rate of extinctions constant per unit mass. The rate of extinctions per unit volume is then proportional to the mass density.

In a perfect vacuum there can be no extinctions, for there is nothing to become extinct. In such a region the net rate of origins is equal to the gross rate. But as matter accumulates in the region a certain constant fraction of it becomes extinct and the net rate is the difference between the *constant* rate of origins and the density dependent rate of extinctions. There must be a specific density for which these rates balance, and the net rate of origins is zero. This is the density that I propose to call the *equilibrium density*.

If the whole universe were at the equilibrium density, its content and extent would be constant. It would neither expand nor contract. But the observed expansion of space proves that the average density for the whole universe is less than the equilibrium value. (This assumes that origins and extinctions of matter are coupled with the expansion and contraction of space. The assumption needs to be justified. Attempts to do so will be found in Part Four and again in Appendix H). If we knew the gross rate of origin of matter per unit volume and time, or if we knew the half-life of mass, we could calculate from McCrea's estimated net increase of 500 atoms of hydrogen per cubic kilometre per year what the equilibrium density is. But as these quantities are not known, we can only say that the equilibrium density is greater than the average density of the whole universe.

By terrestrial standards both are very low values. The average density of our galaxy must, for instance, be well above the equilibrium density. But we must not allow our sense of proportion to be too much influenced by terrestrial standards. We are considering processes in extragalactic space, in most of which the density must be well below the equilibrium value.

During the first stage of growth the cloud depends entirely on new origins for increase of mass. Should extinctions exceed origins the cloud would not grow but dwindle. It follows that the average density of the cloud must be below the equilibrium value during and right up to the end of the first stage of growth.

The central core must, of course, greatly exceed the equilibrium density and so the gas between this core and the top of the crater must be in a very tenuous condition indeed. This situation must continue after the first stage of growth has passed. The top of the crater, now grown to an astronomical ridge or reversal zone, marks the boundary between some- what steep slopes. Hydrogen must be pouring down these as fast as it forms on each side, leaving the slopes near the ridge almost depleted. The inner slope, moreover, the one towards the growing cloud, must be the steeper of the two and must therefore be the more depleted one.

Chapter 15

Quantitative Treatment of Growth of an Extragalactic Cloud

15.1: *The Time between Successive Generations of Nebulae* .
Contrary to common belief it is sometimes easier to talk in mathematics than to talk in English; this is the reason why many scientific papers contain more mathematics than is either necessary or desirable. Contrary to common belief it is also often less precise to do so. For mathematical symbols have a tendency to conceal the physical meaning that they are intended to represent; they sometimes serve as a substitute for the arduous task of deciding what is and what is not relevant; they tend to limit the inquiry to those physical circumstances that are represented by them and are by their nature unable to reveal others that may have a decisive influence on the conclusion that is being sought. It is true that mathematics cannot lie. But it can mislead.

However, the dangers of over-indulgence in formula spinning are avoided if mathematics is treated, wherever possible, as a language into which thoughts may only be translated after they have first been expressed in the language of words. The use of mathematics in this way is indeed disciplinary, helpful, and sometimes indispensable.

It is disciplinary in the sense in which it is always disciplinary to translate a statement from one language into another; this is the best way of revealing shoddy thinking. It is helpful because some statements are of such a nature that it is almost, if not quite, impossible to express them in the language of words; it is reasonable to ask to be spared the ordeal of attempting to do so. It is indispensable because quantitative conclusions can only be reached with the help of algebraic expressions.

Such considerations demand now that the comparatively light task be undertaken of seeking algebraic symbols for some of the concepts that have been discussed in previous chapters. Only the simplest mathematics is needed at this stage of the inquiry. But the field of cosmology that is being studied here is complicated and so far almost unexplored. Once clarity is achieved concerning its broad outline, the remaining problems will lie wholly within the mathematician's domain.

It is easy to see that the interval between the moment when a cloud of one generation begins to form and the moment when one of the next generation, and in adjoining space, does so must average the time that it takes for the linear dimensions of space to double themselves. This interval can be calculated with the help of Hubble's constant H.

Let D be a given distance in space. The rate at which this increases is

$$dD \,/\, dt = HD,$$

the solution of which is

$$D = k \, e^{Ht}$$

Let $D = D_0$ at the moment when one puts $t = 0$ and the above expression becomes

$$D = D_0 \, e^{Ht} \qquad \ldots\ldots \qquad (15a)$$

H is, according to a recent estimate, 185 kilometres per second per megaparsec, which is the same as 5.84×10^{19} kilometres per year per megaparsec and has the dimension of reciprocal time. One megaparsec is 3.084×10^{19} kilometres, from which it follows that

$$H = 1.89 \times 10^{10} \text{ reciprocal years.}$$

When the distance doubles one puts $D \,/\, D_0 = 2$ and obtains $\ln 2 = Ht$, from which $t = 3.66 \times 10^9$ years.

This means that a new nebula begins to form adjacent to a predecessor about every three-and-a-half thousand million years.

This is the time during which the lip of a crater on an astronomical summit goes through the complete evolution of moving outwards, becoming a ridge far from any concentration and flattening sufficiently to attain at its summit the critical gradient at which a new cloud can begin to form.

15.2: *The Rate at which an Extragalactic Cloud Grows in Extent*
It is necessary to consider next whether the calculated time that it takes for the cloud to grow is of the right order of magnitude, lf the cosmological model based on Symmetrical Impermanence is to resemble actuality, the cloud must acquire a significant mass during a time that is appreciably less than 3.66×10^9 years.

For the sake of simplicity let us assume that the cloud is forming on a pass between two nebulae of equal masses m. Although the cloud forms on a summit and not on a pass, this incorrect assumption will serve to illustrate the method of calculation and I hope that it will not introduce a wrong order of magnitude. Let D_{co} be the distance from the pass to one of the nebulae at the moment when the cloud just begins at the very top of the pass, and let D_{cr} be the distance at a moment when the fringe of the cloud reaches a distance r from the pass. Let E_c be the critical potential gradient at which the rate of origins just balances the rate of loss from particles falling away down the slope.

We want to find the time that it takes for the fringe of the cloud to reach a certain fraction of the distance to the nearest galaxy. Let this fraction be α, so that $r = \alpha D_{cr}$

Consider equations (12e) and (12f) . If (12f) were correct, E_c would be reached at distance r at the same time as at every other distance; it would be reached for the separation D_{co}. But the correct equation is (12e) and for this E_c is not reached until the separation has become D_{cr}. This gives two expressions for E namely:

$E_c = -4Gm \, D_{cr} \, r \, / \, (\, D_{cr}{}^2 - r^2 \,)^2$ from (12e) and,

$E_c = - \, 4G \, m \, r \, / \, D_{cr}{}^3$ from (12f)

When one replaces r by αD_{cr} and equates these expressions, one obtains

$- \, 4Gm \, / \, D_{cr}{}^2 \, (\, 1 - \alpha^2 \,)^2 = -4GmD_{cr} \, / \, D_{co}{}^3$ from which,

$$D_{cr} = D_{co} \, (\, 1 - \alpha^2 \,)^{-2/3} \qquad \ldots\ldots \qquad (15b)$$

From this one can calculate that when r is 1 per cent of D_{cr} the ratio $D_{cr} \, / \, D_{co}$ is 1.000067, and when r is 5 per cent of D_{cr} it is 1.00167.

If one puts for D in equation (15a) D_{co} when $t = 0$, one can write

$$D_{cr} = D_{co} e^{Ht}$$

From this it follows that t is 350,000 years when the cloud extends to I per cent of the distance to the nearest nebula, and 8.8 million years when the cloud extends to 5 per cent.

These figures are over-estimates, and probably considerable ones, for they do not allow for the flattening of the slope that is occasioned by the mass of the cloud itself. When this is taken into consideration one must arrive at shorter, and probably substantially shorter, times, particularly for the 5 per cent distance.

The estimate does not allow, either, for any change in the mass m of the neighbouring galaxies. The conclusion has been reached in the last section that this is growing only slowly when cloud formation begins. It will be shown in Appendix B that it may be dwindling later. For the condition that is being considered it must be near the turning point, and so neglect of change of mass is not likely to introduce a big error.

It has been found herewith that, compared with the three-and-a-half thousand million years that are available for the cloud to acquire its final mass, the time that it needs to become rather extensive, though it is also still very tenuous, is quite short.

The orders of magnitude in the model based on Symmetrical Impermanence seem to fit.

15.3: *The Rate of Growth of the Cloud's Domain*
As soon as a cloud has become massive enough to have an appreciable gravitational field of its own, a reversal zone occurs just around the astronomical summit. I have previously described this zone as the lip of a crater. There is some incipient cloud on both sides of this. A portion of this cloud occupies the crater and another portion lies on the outer slopes.

As the mass of the cloud increases, this reversal zone moves outwards; the crater grows in depth and diameter. It continues to do so until the reversal zone coincides with the fringe of the cloud. The first stage of growth then ends and the second stage begins.

We thus have to picture two simultaneous movements, both proceeding radially outwards. One is the movement of the fringe of the cloud, the other that of its reversal zone. This begins like the lip of a crater; at a later stage it envelops the whole cloud, and it extends eventually far out into space, as a range of passes and summits.

The reversal zone can form only after the very tenuous incipient cloud has become fairly extensive, and even then it is no more than a small shell right at the centre of the cloud. To extend eventually beyond the cloud it must grow outwards more quickly than the fringe does. Some simple mathematics shows that this happens.

In practice the cloud's shape must be very irregular, but for the sake of simplicity a spherical shape will be assumed here. This simplification may lead to quantities that would need a fairly large correction factor. But it suffices to give an idea of the relative movements of the fringe and the boundary.

Let at any moment the distance from the centre of the cloud to the reversal zone be D_1 and let the distance from the neighbouring galaxy to the same point on the reversal zone be D_2. Let at the same moment the mass of the cloud within its domain be m_1 and the mass within the domain of the neighbouring galaxy be m_2. By the inverse square law

$$(D_1 / D_2)^2 = m_1 / m_2 \qquad \text{......} \qquad (15c)$$

Let the distance from the centre of the cloud to its fringe be r and let the further simplification be introduced that the mass of the cloud is proportional to its volume. One can then write

$$m_1 = a\, r^3 \qquad \text{......} \qquad (15d)$$

where a depends on the shape and average density of the cloud. Combining these two equations gives

$$D_1 = D_2\, (a / m_2)^{1/2}\, r^{3/2} \qquad \text{......} \qquad (15e)$$

If D_2, a and m_2 have given values, D_1 is seen from this equation to increase more rapidly than r. This is perhaps easier to appreciate if one replaces equation (15e) by

$$D_1 / r = \text{constant x } r^{1/2} \qquad \text{......} \qquad (15f)$$

This expression shows that D_1 is smaller than r when r is small and larger than r when r is large. There is one particular value for r when it equals D_1. This is the value for which the fringe of the cloud coincides with the reversal zone.

Too much importance must not be given to the above equations. They do little more than to provide reassurance that in the model based on Symmetrical Impermanence the time must arrive rather soon in the history of an extragalactic cloud when it ceases to grow in extent, when indeed it begins to shrink under its own gravitational field, and when at the same time it grows in mass by capture of hydrogen from without.

Chapter 16

Observation, Inference And Speculation

It will be convenient if I summarize very briefly the conclusions that have just been reached. They are as follows.

One should expect a cosmological model based on Symmetrical Impermanence to contain some extensive clouds of very tenuous gas. In an expanding universe these would begin to form as soon as the distance between adjacent galaxies exceeded a certain critical value. At first these clouds would grow rapidly in extent, while they still remained so tenuous that the gas between stars within our galaxy would have to be called dense in comparison. But when these extragalactic clouds had acquired a substantial gravitational field of their own, they would capture hydrogen from their surroundings and would also shrink under the influence of their own gravitational fields. Hence they would become more dense. But in spite of their smaller volume they would also become more massive, for loss from matter falhng out of the cloud would cease and be replaced by gain from the matter that fell into it.

These clouds would bear but little resemblance to the spiral nebulae. Nor will the closer study of their formation that is to follow shortly improve the resemblance. On the contrary. It will be found that the incipient clouds must differ from spiral nebulae quite significantly in shape, in size, in the way matter within them is distributed, in their own intrinsic motion.

Can it then be inferred that the cloud will eventually develop into a spiral nebula, into a galaxy similar to those with which astronomers are familiar? If one cannot infer this, the cosmological model that is based on Symmetrical Impermanence does not resemble actuality and this basic hypothesis must be abandoned and with it the even more basic Principle of Minimum Assumption. Much depends on whether one can infer from established knowledge and this principle that the extragalactic cloud can develop into a familiar spiral nebula.

For this reason it now becomes necessary to consider in detail what one should expect to happen to an extragalactic cloud after it has begun to form on an astronomical summit. But such consideration can with ad- vantage be preceded by an appreciation of the task that it involves, and this will be most easily obtained, be it at the cost of some repetition, if I recall what has been said in Chapter 2 about scientific method.

To infer from established knowledge what happens to an extragalactic cloud is comparable to what engineers do when they infer the performance of machines that have been designed but not yet made machines that may embody entirely new principles and that may be quite different from anything that has been made before.

To draw such inferences taxes the engineer's powers of logical reasoning considerably. It calls for great care lest some decisive factor be overlooked; it may require the use of some new mathematical tool; the design may have to be preceded by much experimental research of which the purpose is to extend the range of established knowledge. But in engineering the new machine does not have to be manufactured and tried out before anything is known of its performance. The designer knows, of course, that some mistake or unforeseen circumstance may falsify his predictions, so his performance figures include margins, technically known as tolerances. When there is some doubt, a prototype may precede mass production. But these precautions do not imply that the engineer lacks all faith in his own powers of calculation and logical inference; and in the great majority of cases experience shows a close agreement between inference and eventual performance.

Amateur scientists and philosophers often fail to appreciate this. The technique of inference, which is a commonplace in engineering, appears to them as unpardonable speculation. A consequence of this is that both scientific and philosophical conclusions are too often assessed by the wrong criterion. When a conclusion is arrived at by a process of inference, the proper course is to examine the reasoning that has led to the conclusion. This may have been faulty and can then be exposed by better reasoning. But the history of science abounds in occasions when this obvious duty has been neglected. For one critic of Darwin who examined his reasoning in the early days of evolutionary theory a hundred dismissed the theory by quite different, and far less sound, criteria.

There are sundry reasons for this and one is man's natural intellectual indolence. The detection of faulty reasoning requires a substantial mental effort, while other criteria can be applied with little or none. One of these false criteria is applied when the result of inference from established knowledge is dismissed as 'mere speculation'. If this happens more often than one should expect, it is because the difference between inference and speculation tends to be obscured.

Confusion sometimes arises from giving a false status to observation. There are schools of philosophy in which the validity of any statements about unobservables is denied. These schools find ready adherents, for the notion

102

seems superficially plausible that one cannot say anything valid about something that one has not and cannot observe. However, the plausibility evaporates when one remembers the everyday work of scientists and engineers.

A machine that has never been made is certainly unobservable at the time when the designer makes statements about its performance, and so the philosophers of this school would have to say that such statements could not be justified scientifically. These philosophers advocate a kind of empiricism that would make the engineer's task impossible. It is fortunate that most physicists, and all engineers, know better than to practise what is sometimes preached to them.

Cosmology is typical, but by no means unique, among the fields of study in which one cannot afford to listen to these schools. In cosmology inferences have to be drawn about systems that are just as unobservable as a machine that has not yet got past the blueprint stage. Predictions in this field require the same disciplined care as in engineering; if this care is exercised, they are just as valid; they are subject to the same kind of un- certainty. As often happens in engineering, moreover, inferences cannot be checked with the help of small-scale models.

All this applies to the present theme. We are here trying to draw inferences about the behaviour of masses of gas so enormous that their movement is largely governed by their own gravitational fields. Such masses are not at the disposal of the laboratory technologist. The telescope tells us a little about them, but by no means all that we need to know. If we are to extend our knowledge in the field of cosmology, we must adopt the engineer's technique of inference. There is no alternative and those who deprecate inference do not suggest one. One must either use this technique with the thoroughness and care of the machine designer, or abandon all hope of gaming a better insight into cosmological processes.

Chapter 17

The Forces That Operate in Extragalactic Space

The technique of inference that has been discussed in the preceding chapter calls for a survey of all the forces that could conceivably have an influence on the course of events. The following list seems to be exhaustive: gravitational forces, centrifugal forces, forces of impact, pressure of radiation, electrostatic forces, magnetic forces. But a moment's thought shows that some of these cannot materially influence the shape or behaviour of an extragalactic cloud.

Electrostatic and magnetic forces, for instance, can have a big influence on small-scale events. They can cause turbulence in a gas. There is little doubt that they play a large part in terrestrial and solar phenomena of various kinds. But the known processes by which such forces are produced set a limit to the distance over which they act.

An electrostatic force occurs only when electric charges have been separated, and the region within which the force acts is the region between the places to which the charges have been removed. If an electrostatic field is to cause hydrogen to move through a great distance, the field must extend over at least the same distance. For this to happen charges must have been torn asunder and removed to the extremities of the field. This must have been done against the force of their mutual attraction. So the charges must have been subjected for a long period of time to a unidirectional force before the field could be produced. In the process vast quantities of energy must have been employed. Now from established knowledge, and without the help of some quite fantastic *ad hoc* hypothesis, it is not possible to infer a process that would separate charges by the requisite distance.

It is by the same reasoning that one is obliged to dismiss magnetic forces from the list of those that could cause the unidirectional movement of matter over any distance that was an appreciable fraction of the diameter of a galaxy. I am aware that at the time when this is being written the hypothesis has considerable backing that large-scale movements within galaxies are controlled by magnetic forces. It cannot be proved that they are not, but I should be more inclined to accept this hypothesis if its authors showed more awareness of the difficulty there is in finding a cause of the hypothetical magnetic forces. Magnetic forces are known only as the result of the movement of electric charges. They operate only over such distances as may form the limits of an electrostatic field. It is not so difficult to believe that electric charges must circle the earth and cause opposite places on it to have, respectively, north and south polarity. But one would have to go a very long way in search of an hypothesis by which charges would encircle a noticeable part of a galaxy and give it a particular magnetic polarity.

That external pressure of radiation has to be ruled out has been explained already in Chapter 11. Such pressure, it will be remembered, must operate generally in the same direction as gravitation and diminish according to the same inverse square law. But its intensity must everywhere be negligibly small in comparison with the intensity of the gravitational field.

Internal pressure of radiation must, however, have a significant effect on any body hot enough to generate much radiant energy in its interior. It is known to be so for the stars and must also be so for the hot central core of a spiral nebula. However, we are now considering the early cloud and we have not discovered any reason why this should be hot. When a cloud begins to form on an astronomical summit, its component molecules can be moving only under the influence of the faint gravitational fields in which they find themselves. Their velocity must be so slow to begin with that the cloud must be quite near the absolute zero of temperature. It should not be overlooked, however, that internal pressure of radiation on the very large core of the cloud may have a significant effect at some later stage.

Thus inference (as distinct from speculation) only permits three kinds of force for the construction of a cosmological model based on Symmetrical Impermanence. They are: gravitational forces, forces of impact, and centrifugal forces. If these alone do not serve for inferring a model that resembles actuality. Symmetrical Impermanence will have to be replaced by some other basic hypothesis.

Chapter 18

The Early Shape and Structure of An Extragalactic Cloud

18.1: *Variations in Density*

Though very tenuous everywhere, the cloud cannot have a uniform density throughout. Let us consider in broad outline how the density must vary from the centre outwards.

As in the earth's atmosphere, any gas in the cloud that bears the weight of higher layers must be compressed. So the greatest density must be at the centre of the cloud.

A space traveller who moved outwards from this centre would be rising against the cloud's gravitational field and would be entering an increasingly rarified atmosphere as he did so. After a while he would find himself climbing up the inner slope of the ridge that formed the lip of the crater around the astronomical summit. This ridge or reversal zone, it will be remembered, is moving outwards as the cloud grows in mass. A little earlier the place where the space traveller is at the moment was on an outward slope. Gas was falling away from it into further space and is now falling in the reverse direction towards the core. So it is safe to assume that none of the gas that was there originally can have remained. The only gas that the space traveller encounters must be new particles that have originated but recently. On the inner slope of the crater the gas density must be as near to zero as it can be anywhere in space.

On crossing the ridge the space traveller reaches a very gentle gradient that slopes downwards away from the centre of the cloud. During the first stage of growth gas is originating in this region faster than it falls down the slope and the traveller enters once again into a region that is less depleted. The density is here much as it was on the summit at the still earlier stage, when the cloud first began to form as a small sphere.

As the ridge moves steadily outwards, the picture changes. What has just been on the top of the ridge finds itself on the inner slope and falls backwards on to the core in the hollow of the crater. The change goes on until the ridge has reached the fringe of the cloud and the first stage of growth has been completed.

Thus there must be a period during the first stage of growth when the cloud has a central core, which may already be fairly dense at its centre and is very tenuous in its outer layers. Around this there must be a region where there would be no gas whatever if it were not for new origins within the region. It is easy to show from the inverse square law that the potential gradient must be relatively steep in this region and so the new origins fall quickly out of it.

Beyond the region of the near-perfect vacuum there must be more cloud. It must extend a long way out into space and be very tenuous; for its density must be below the critical value for cloud formation. The potential gradient becomes steeper as one proceeds from the crater outwards and so the gas density must also diminish with distance from the core.

It will be found later that the large vacuous region around the core plays an important part in the evolution of galaxies and so it needs to have a name. I propose to call it the 'depleted region'.

18.2: *Rotation of the Core*
Stars and galaxies are known to rotate, but it is far from easy to say why. I doubt whether it is generally appreciated how difficult it is to find an explanation. Any explanation that is found must conform to established principles of mechanics and it is desirable that it should not require the invention of *ad hoc* hypotheses. I doubt whether a satisfying explanation can be based at all on accepted notions about gravitation. The difficulty, together with a tentative solution of the problem, will be discussed in Chapter 28. It cannot be usefully discussed until after a discussion of the nature of gravitation has been undertaken. In the meantime it must suffice to note the bare fact that galaxies are observed to rotate.

18.3. *The Physics of the Core*
It would be a mistake to seek much resemblance between the core of the extragalactic cloud and a star. Apart from the facts that both spin and consist mostly of hydrogen, they can have little in common.

The core is very new at the time we are considering, i.e. before the reversal zone reaches the rim of the cloud. The time that it takes for the zone to traverse to this has been found to be probably no more than some hundreds of thousands of years. The time taken to travel far enough to leave a core behind must then be reckoned, perhaps, in tens of thousands. The extent of the core is enormous; it could contain many millions of stars. So the journey travelled by those molecules that contribute to the shrinking is a long one. After a hundred thousand years or so the core must still be large enough to be very tenuous. It can have no property that one could think of as rigidity. Collisions between molecules must be rare by terrestrial standards. The quantity of heat generated per unit volume by collisions must be small.

In such a core one should not expect the conditions to exist that cause the synthesis of hydrogen into helium; this can only happen with pressures and temperatures such as occur inside the sun. The heat generated inside the core does not seem to have any other source at this time than the gravitational potential energy of the molecules of hydrogen, which is very small compared with the energy released by the synthesis of helium. The temperature in the core can therefore only be the result of the small amount of heat generated by collisions. The core must be a very cold body.

This must have a significant effect on the rate of shrinking. In a star the helium synthesis produces intense internal radiation. Its pressure is so great that it balances the star's own gravitational field. In other words, pressure of radiation supports the weight of the outer shell and prevents it from falling towards the centre. When this pressure ceases, as sometimes happens, the star collapses rapidly and with a great increase in the rate of spin.

As it does not seem possible to postulate the restraint of internal pressure of radiation for the core during its early life, one must assume that the penetration of molecules into the core is restricted only by collisions; and as, in view of the core's tenuousness, these are rare, penetration must be deep. The rate of shrinking must therefore be great.

18.4: *Retardation of Spin*
From this one might at first be inclined to infer that the angular velocity of the core would rise to the point where centrifugal force balanced gravitation. If this were to happen, there would be no shrinking. But there is also a retarding influence that we have to consider.

While the cloud is shrinking and becoming more dense by reduction in its volume, it is also becoming more dense by the acquisition of new matter, and this from two distinct sources.

The first of these is new origins within the region occupied by the cloud. The new particles would appear to an observer to be at rest relative to a given frame of reference. The existing particles must collide with the new ones and entrain them. Hence the new particles come to participate in the general rotatory motion. The angular momentum of the core remains constant, but the angular velocity becomes less as the mass increases.

The other source of new matter is the gas that finds itself within the reversal zone as this moves outwards. This gas, which has hitherto been falling away from the centre of the cloud, now falls towards it across the depleted region. It has to fall a long way across this steep part of the inner slope where, as has been mentioned above, there is a high vacuum. During this falling it acquires a substantial momentum and must penetrate deep into the core. Like the originating matter, it is entrained and causes retardation.

It would be interesting to know more about the physics of the core. How does its density vary from the centre, where the only force is the weight of the gas around it, to the periphery, where centrifugal force opposes the weight? How does the angular velocity vary with time? Does it increase at first up to a maximum and then decrease as the retarding effect of additional mass becomes more pronounced? To what extent is the core flattened by rotation? Is it disc-shaped at first, and does it become spherical later when pressure of radiation from helium synthesis blows out, as it were? When, if at all, does helium synthesis begin? But such questions must be put to various specialists in astrophysics. They are beyond the scope of this elementary study.

18.5: *The Spokes* It has been pointed out in Chapter 10, entitled 'The Astronomical Landscape', that the potential gradient around an astronomical summit cannot be at all uniform. There must be shoulders and valleys and these must stretch out into space in several directions in all three dimensions The shoulders descend gently for a long while and eventually reach those places that have been called astronomical passes. The valleys descend more steeply and end in the deep wells that represent the neighbouring galaxies The number of pronounced shoulders, it will be remembered, cannot be great. A typical number is likely to be six; and these will spread out from the summit like rods pushed through the faces of a cube. If all surrounding galaxies were arranged in a regular cubic pattern, the six astronomical shoulders would be at right-angles to each other.

The cloud can only form where the potential gradient is below this critical value. This must occur along a gentle shoulder at a time when the valley next to it is so steep that no gas can accumulate there. The cloud cannot therefore have the form of a simple sphere. Its shape must be very different. Hydrogen must be accumulating on a shoulder while it is rapidly pouring down the side into the deeper valley below. It must also be pouring down this towards the neighbouring galaxy.

This shows that the inner cloud must, during its early stages, be surrounded by a collection of spokes that extend away from the summit in all directions and far out into space. The spokes must be roughly cylindrical and must become more tenuous towards their tips, where the potential gradient is greatest.

As space expands, the landscape is flattening in all directions and so the spokes must grow in diameter as well as in length. They must also acquire a gravitational field of their own. In other words, they must carve out shallow troughs for themselves along the shoulders. They may be thought of as lying in these and attracting hydrogen towards their axes, Hydrogen in the outer regions of a spoke must fall inwards towards its axis and thereby collide with hydrogen that is already there. It will give rise to some turbulence. But I should not like to claim that the turbulence would be enough to have any significant effect. To all intents and purposes one may perhaps regard the spokes like inert slugs lying motionless along their supporting shoulders.

Though I have called them spokes, they are not spatially arranged like the spokes of a wheel. They are not all in one plane but point in each direction of three dimensional space; and they are not connected to a central hub. The wide depleted region separates them from the inner core of the cloud.

The picture that we have been led to infer differs substantially, as I have said already, from that presented by the spiral nebulae.

Chapter 19

Cloud Into Nebula

19.1: *Whence the Semblance of Rigidity?*
Our galaxy rotates about its axis. So do all the spiral nebulae. Their great arms wheel in unison like platoons of soldiers on the parade ground, though not with the same angular velocity as that of the core. This presents us with a problem that has, I am afraid, not been appreciated. The nebulae are mixtures of a diffuse gas and stars. The distance between a star and its nearest neighbours is usually many light years. How is it that this flimsy mixture turns around a common axis as though it were a viscous fluid?

It is surprising that there has not been a more assiduous search for an answer to this question. It seems usually to be ignored, perhaps because it has been regarded as unanswerable; it can hardly be thought that the answer is known already. So far as I know, only two explanations have ever yet been offered and less tenable ones are hard to imagine.

One is that magnetic forces act between the component parts and are powerful enough to lend the property of rigidity to the tenuous structure. I have already shown in Chapter 17 that an hypothesis that attributes large scale unidirectional movement to magnetic forces does not stand the test of quantitative thinking. The other explanation so far given is that tides raised by the stars on each other bring them all into line. I do not think there is any likelihood that astronomers will have seriously entertained this strange theory, but others may not feel qualified to judge it, and so it is worthwhile to point out that, like the hypothesis of magnetic forces, it fails badly by the test of quantitative thinking.

It is true enough that tides do impart a small measure of rigidity to an astronomical system. They tend to bring parts that are moving relatively to each other into unison. The mechanism by which this happens is rather complicated but it is well-known to experts and so does not need elaboration. It will suffice to mention that tides have caused the moon always to present the same face to the earth, as if the two were connected by a rigid rod. Tides around our estuaries are also slowing down the earth's rotation, and if they persist for long enough, they will similarly cause the earth always to present the same face to the sun.

Tides raised by the sun, the planets, and their satellites on each other are also

111

tending to cause the equator and the ecliptic for each of these bodies to coincide. They are tending to cause all the planets to rotate with the same angular velocity about the sun, just as the arms of the spiral nebulae rotate with the same angular velocity about a common axis.

But tides within the solar system are as yet far from achieving this; they have not yet caused any of the planets, except the smallest and nearest to the sun, always to present the same face to the sun. Still less have they brought all the planets into a common ecliptic or caused the year to last the same time for all of them. And yet the tides within the solar system are powerful compared with those that can be raised by one star on another one several light years distant. If tides have not caused all planets to have the same solar year, they can certainly not cause all fixed stars to have the same galactic year. The notion that tides could bring stars into alignment is as absurd as thinking that a fly could deflect a cannon ball.

19.2: *Uniform Rotation as the Consequence of Impacts*
I have said in Section 18.2 that it is difficult to understand by what process a celestial body, be it a star or a galaxy, can acquire the observed angular momentum. But whatever the explanation may be, I know of only one way by which uniform rotation could be imparted to all the com- ponents of so flimsy a structure as a cloud of gas, and this is by successive impacts of molecule on molecule. But as appropriate impacts are not being made on the stars now, they must have been made on the substance forming the stars at some time in the past.

Rotation can only be imparted to the molecules that are captured from outside because these have sufficient momentum to fall, not only *on to*, but *into* the core. If the captured molecules were to rest on the surface they would not be entrained. That a spherically shaped cloud such as the core will rotate as a whole is only to be expected provided always that there is a source of angular momentum and that the core grows from the centre outwards, and this is ensured by the outward movement of the reversal zone and is accentuated by the small difference between equations (12e) and (12f) in Section 12.4, which prevents the cloud from forming simultaneously over a large volume.

19.3: *The Part Played by the Spokes*
Let us consider what the cloud looks like at the moment when the reversal zone has reached the base of the spokes. There is a central core, which is becoming denser from molecules that fall into it as well as from shrinking - two distinct causes. Around this core there is a potential gradient that slopes towards the centre and down which hydrogen is falling. It only contains matter that is, as it were, in transit; and has, because of its extreme tenuousness, been called the depleted region. This is bounded by the expanding reversal zone and beyond this lie the spokes. It is obvious that

there can be no collisions between molecules in the rotating core and those in the motionless spokes; there is a great gap between them.

The structure so far inferred does not resemble the spiral nebulae. To become like them it will have to change greatly. I think that it must do so and in the following manner.

When the reversal zone reaches the base of the spokes, masses of hydrogen outside this zone pour into it in the few places where there are spokes. We have to picture half a dozen or so well separated places from which hydrogen enters the domain of the core. It is as though the core were surrounded by a far distant outer shell with this number of openings through which gas is pouring into the interior. Gas continues to do so as long as, first the base of each spoke and then successively more distant parts, are traversed by the reversal zone and find themselves attracted towards the core.

Here it is necessary, however, to guard against exaggerating the analogy to a terrestrial system. Gas would only pour into a steel vessel if the pressure was greater outside than inside; and if this happened the gas would, after entry, diffuse almost instantly throughout the whole volume; the pressure and density would even out very quickly. But in our astronomical system it is gravity and not pressure that causes the substance of the spokes to pour inwards. The pressure and density considered in a radial direction are far from uniform. Under its own weight the core is dense at the centre and tenuous at the surface. Above this surface there is the depleted region through which the gas is falling.

While falling across this region, molecules from the spokes are being accelerated. By the time they reach the surface of the core they must have attained a substantial momentum and must penetrate deeply. They are entrained by molecules that already have a tangential velocity in the direction of rotation.

This shows that the process that introduces rotation is different from what one might at first sight suppose. The substance of which the spokes are formed does not take part in any rotatory movement at the place far out in space where they form; it only does so after it has fallen a long way down into the rotating core.

19.4: *A Tentative Explanation of the Spiral Shape of the Arms*
In a structure built to terrestrial dimensions the substance of each spoke would quickly spread out all around the core. But we are here concerned with a structure so vast that light itself takes a very long time to traverse a small fraction of its span. The rate at which gases diffuse is very slow compared with the velocity of propagation of light, even at atmospheric pressure. In the extragalactic cloud the collisions by which diffusion is effected are

comparatively rare, so the diffusion rate is correspondingly slow. For practical purposes the effect of diffusion on the bulk movement of gas must be negligible compared with the effects of gravitation and centrifugal force.

This means that to all intents and purposes the substance of the spokes stays, after falling, where it has penetrated, deep within the body of the core. Here, as we have already seen, there is a local thickening.

While these falling molecules are being entrained they slow down those that entrain them. Each place into which the substance of a spoke is falling is therefore not only denser than the rest of the core but also rotating more slowly. In front of each thickening the gas of the core moves away, leaving the region there more tenuous. Behind the local thickening, on the other hand, the faster molecules pile up, causing the thickening to extend backwards. This thickening must therefore have a well-defined leading edge, in front of which there is a near vacuum, and behind which the thickening extends for some distance and tails off slowly.

As time goes on more substance from the spokes pours into the core; but as the core is moving forwards the later substance always reaches it at a place further back. There is, in other words, an increasing angular displacement between the leading edge of the thickening and the place at which each successive part of the spoke arrives, and as the further gas falls on to piled-up substance it penetrates less deeply. It consequently finds itself in a place that is both tangentially behind the previous part and radially further from the centre. Thus does the leading edge of the thickening acquire a spiral form.

The spokes must not, it has now become clear, be thought of as the embryonic arms of the nebula. The spokes do not form where the arms will form but much further from the centre of the cloud; and they have quite a different shape. The spokes are really a motionless store from which the rotating core acquires the substance that is to be fashioned into the spiral arms. But the spokes provide only a part of the arms. The remainder consists of gas from the core itself, which piles up behind the thickening where substance from the spokes is falling into the cloud.

The comparatively small dense ball that is observed at the centre of the spiral nebula must not, moreover, be identified with the part of the early cloud that I have called its core. This latter extends over and beyond the whole volume that will eventually be occupied by the nebula. The ball at the centre is but the inner compressed part of the core.

The function of a store will be served by each spoke whether it lies in the plane of rotation or not. Should the axis of the core be so inclined that four spokes happen to be in this plane and two along the axis, the resulting nebula will have about four spiral arms, rarely more, and they will be of about equal sizes and equally spaced. The other two spokes will probably be absorbed into the central ball without forming arms. But with a different inclination of the axis there may be a different number of arms, and these will be less uniform both in size and position. They will, however, always occur in the plane of rotation and be flattened out by centrifugal force.

19.5. *The Extent of the Cloud on Completion of the First Stage*

We are now able to form some rough and tentative conclusion concerning the span of the cloud at the time when it reaches its maximum extent and the first stage of growth is completed. Some time before this moment the lower parts of the spokes have been falling down into the rotating core. Now, at the end of the first stage, it is the very tips that do so.

The core has been shrinking and so the tips of the spokes must have a very long way to fall; one must consequently conclude that the radius of the core is small compared with the distance from the centre to the tip of a spoke. But it has been found that the arms of the spiral nebulae must have been formed wholly within the rotating core, so this core cannot be less voluminous than the nebula that will eventually appear. Indeed it may be more voluminous; for it probably goes on shrinking long after the spiral arms have been formed.

The reason for this supposition is that the density of the cloud must increase by a large factor during the second stage of growth. Consequently its angular velocity must decrease and with this the distance from the centre at which gravitational and centrifugal forces balance. From these considerations we have to conclude that the span across the spokes reaches a value several times greater than the diameter of the nebula that will eventually result.

There is no apparent reason why the shape of the spiral arms should change very rapidly. The substance of which they are formed is subjected to two opposed radial forces, namely gravitation towards the centre and centrifugal force away from the centre. One should expect a balance between these to be reached in the course of time. For there is some reason to believe that the mass of the nebula will eventually fluctuate around a limiting value, as will be explained in Appendix B.

It seems possible that the spiral arms may become narrower and lose their clear outline with time. Each has its own gravitational field and one might expect substance at its fringe to be attracted in a tangential direction towards this. Given time enough one should expect further departures from the original shape and a less and less regular structure. The structureless nebulae might be degenerate spiral ones. But elliptical nebulae would still remain unexplained. They may even make it necessary for the theory presented here to be revised.

19.6 *Why Stars?*

In one important feature the model that has just been inferred is quite unlike actuality. It consists entirely of diffuse hydrogen, whereas nebulae consist of a mixture of diffuse hydrogen and stars. Must the model be abandoned for this reason, or can one infer stars, too, only on the basis of the Hypothesis of Symmetrical Impermanence?

This question worried me for some thirty years and restrained me from previously publishing a cosmological model based on Symmetrical Impermanence. But the question has got to be faced seriously and with a proper sense of scientific responsibility.

Rather surprisingly this does not seem to have been done yet any more than the question seems to have been asked seriously why the flimsy nebulae rotate as though they were viscous structures. To both questions only untenable answers seem to have been suggested.

One rather surprising explanation that has been offered to account for the occurrence of discrete stars is that, from considerations of probability, one might expect some lack of uniformity in any extragalactic cloud. There would be regions of higher and regions of lower density. The regions of higher density would have centres of gravity of their own and each would attract hydrogen to itself and away from the more powerful, but also more distant, centre of gravity of the whole cloud.

This theory is probably not seriously entertained by any astronomer or astrophysicist, but others may accept it and so it is worthwhile to men- tion that it is as untenable as the above-mentioned theories about the effect of magnetic forces or tides in giving rigidity to the nebulae. Like those, it does not pass the test of quantitative criticism.

This theory postulates local, or parochial, centres of gravity in addition to the common centre for the whole cloud. It claims that the parochial centres are powerful enough to compete with the common centre. For this to happen the potential gradient near each parochial centre must be reversed, the slope must be away from the common centre and towards the parochial one. But a moment's thought shows that, according to the accepted views about gravitation, the density of the gas around the parochial centre would then have to be enormous.

The following calculation would not be precise if an actual situation had to be evaluated, but is near enough to the truth to convey a correct sense of proportion.

Consider a point at a short distance D_1 from the parochial centre and at a large distance D_2 from the common one. Let the density around the parochial centre be σ_1 and the average density for the whole cloud have the lesser value σ_2. For the sake of simplicity let the mathematics be applied to spherical regions around the parochial centres. The masses that are in competition at this point are then

$m_1 = (4/3) \pi D_1^3 \sigma_1$ and $m_2 = (4/3) \pi D_2^3 \sigma_2$

For the gradient to reverse we must have

$(D_1 / D_2)^2 = m_1 / m_2 = (D_1 / D_2)^3 (\sigma_1 / \sigma_2)$

It follows from this that

$\sigma_1 / \sigma_2 = D_2 / D_1$ ...... (19a)

The number of stars in a galaxy is perhaps 10^9 or more. This means that each star must draw its substance from a region that is but a tiny fraction of the whole volume. Hence D_1, the radius of each of these small volumes, must be very small indeed compared with the average distance from the parochial centre to the common centre. It follows from equation (19a) that the local density around the parochial centre must also be an enormously high multiple of the average density of the whole nebula if this local centre is to be effective in competition with the common centre.

On probability considerations, however, one is entitled to expect only small local departures from the average density. Once again quantitative thinking defeats a theory of superficial attractiveness.

Those who find this theory acceptable have probably been misled by the analogy to a terrestrial cloud of water vapour. This breaks up into droplets of water, each very small in comparison with the whole cloud. Hence, the argument by analogy may have run, an extragalactic cloud would break up into droplets each as big as a star, but small in comparison with the whole cloud. But one must not forget that this is no analogy.

The gravitational field of a terrestrial cloud is negligible; there is no tendency for molecules of water vapour to move towards a common centre of force. The forces that cause droplets to form are, moreover, not gravitational; they are electrostatic and free from competition from a common centre. Compared with gravitational forces they are also, per unit mass, very powerful. The occurrence of these parochial centres of attraction is well understood and does not depend on variations in local gas density.

Nevertheless, it does not seem possible to account for the condensation of

117

discrete stars out of a diffuse cloud unless one can postulate parochial centres towards which molecules of hydrogen are attracted with a force sufficient to overcome the gravitational force that the whole cloud exerts. So the question arises whether such parochial centres can occur in a gas that is virtually of uniform density.

At the moment it must seem as though such powerful centres could only be postulated on the basis of some *ad hoc* hypothesis. But it will appear in due course that this is not necessary. Such centres do follow, surprising though it may seem at present, as a logical consequence of the Hypothesis of Symmetrical Impermanence.

Before this can be made clear, however, another inference will have to be drawn from Symmetrical Impermanence. It concerns gravitation and will be the object of Part Four of this essay.

Chapter 20

Evidence From the Radio-telescope?

Inferences, as I have pointed out already, may generate false conclusions either because they are based on false premises or because they have proceeded by a faulty process of reasoning. Hence one seeks, wherever possible, to test inference by observation. Does any observation provide evidence for the story that I have just unfolded about the early life history of the nebulae?

Insofar as what is provides evidence for what has been, there is plenty of it. It has been shown in the last few chapters how those details that we observe in the nebulae are just what one should expect to observe if my theory is true. But the theory that I have presented may not be the only one that could serve to explain what is observed. It frequently happens that rival explanations account for the same facts. There were in the past, for instance, rival explanations for the observation that days and nights succeed each other. One was that the earth rotates about its axis, the other that the sun and stars rotate about the earth.

On such occasions one has to compare the two explanations and test each for its logical consistency and its compatibility with established knowledge. So it may be here. Perhaps it will be possible to find a different evolutionary history for the nebulae, one that will equally well explain their structure and rotation. True, this has not yet happened; my theory cannot yet be rejected in favour of an alternative; it can only be rejected in favour of nothing. It is, however, to be hoped that the search for alternative explanations will be undertaken in due course, and with appropriate persistence. Meanwhile one would like, if possible, to be able to observe a spiral nebula in the making.

Is there any hope of this? Let us consider where one should look for such an event.

In my model new nebulae begin to form on all sides of an existing nebula as soon as this is about 366×10^9 years old. There must always be some nebulae that have recently reached this age and so there must be some others, close to them, that are just now going through the early stage of their development. It is in such places, if anywhere, that nascent galaxies could be observed.

We cannot, however, expect such very young nebulae to be luminous. We know from our own galaxy that luminosity comes from the stars in it and not from the interstellar gas. We also know that its cause is the helium synthesis that occurs within them. But this does not begin until the temperature and density in the stars are very high, and it must take a long time for these necessary conditions to be reached. When they are reached, moreover, the rate of helium synthesis is very slow at first. It only gets properly under way when a substantial amount of nuclear energy has already been converted into heat. So the helium synthesis has to proceed for some time before the surface brightness becomes conspicuous. Hence it does not seem to be probable that the amount of energy converted into radiation per unit volume can be enough to make a star become luminous until the nebula has existed for a substantial portion of the 366×10^9 years that represent the time from one generation to the next. But the happenings that determine the shape and structure of the nebula occupy, according to my account, only a minute fraction of this period.

If one cannot detect young nebulae by their luminosity, can one hope to detect them by their opaqueness? I do not think it likely. Much of the far denser gas in our own galaxy is fairly transparent. Only certain patches are opaque. So the amount of light that a very young nebula intercepts is probably quite small. The nebula might reveal its presence as a darkish speck in the sky, but this is doubtful.

There remains, however, one further possibility. This is the radio-telescope. Something very violent, not to say cataclysmic, is happening at about the time when the first stage of growth is being completed: vast quantities of hydrogen are pouring out of the spokes into the core.

We have seen that in this process the gas falls from a great height. In falling a large amount of potential energy is being converted into kinetic energy. The gas in the place where the core thickens must be in a state of turmoil. Collisions between molecules must be vigorous and must emit some radiation. Its wavelength can hardly be in the optical part of the spectrum, but may well be in the part to which the radio-telescope responds. Although the rate of release of energy per unit volume must be small, the volume from which it is released is very great, larger than that of a typical nebula. There thus seems to be quite a good prospect that evidence of the process by which the spiral arms are formed would be picked up in a radio-telescope.

A few things can be predicted about this particular radiation and should help to identify it.

Firstly, the radiation must come from places that are about midway between existing nebulae, i.e. where optical telescopes show nothing.

Secondly, this radiation is only emitted during the comparatively short time while the spokes are discharging their hydrogen into the core and the substance of the core is piling up behind the resultant thickening. This is only a very small fraction of the time that elapses between one generation of nebulae and the next. Perhaps the fraction is 10^{-5}, perhaps less. The number of radio messages of this particular type must therefore be very small in comparison with the number of observed nebulae.

Thirdly, it may be possible to calculate the wave-length of the radiation that would result from violent collisions between molecules of hydrogen in a very tenuous gas. The radiation to look for is of this wave-length.

Fourthly, a new generation of nebulae must begin on every side of an existing one when this is about 366×10^9 years old. This shows that the particular kind of radio message that is to be looked for will hardly occur here and there in isolated places, but tend to occur in clusters surrounding an existing nebula.

PART IV

GRAVITATION

Chapter 21

Inert, Gravitational and Attracting Mass

21.1: *Conceptual Distinctions*
It has been pointed out by others that the word 'mass' has three distinct meanings. Arthur Zinzen, for instance, discussing these in his *Praktische Naturphilosophie* (1953, Westkulturverlag), quotes G. Hamel, who had previously distinguished between these meanings in 1927 (*Hand- buch der Physik*, Vol. 5, Julius Springer). The three meanings are named by him as follows:

(1) Inert mass
(2) Attracted (weight) mass
(3) Attracting mass

The first two of these meanings have also been discussed by Einstein in his *The Meaning of Relativity*. He used the respective words 'inert' and 'gravitational' mass. I shall use the same ones here, but Einstein had no need to discuss the third meaning and so he gave no name to it. I shall therefore use Hamel's name 'attracting mass'.

Before its significance is discussed it has to be made clear why mass in its first two meanings is important in general relativity. For this purpose it is best to use Einstein's own, very precise, words. On page 55 of *The Meaning of Relativity*, Sixth Edition, he said:

'The ratio of the masses of two bodies is defined in mechanics in two ways which differ from each other fundamentally; in the first place, as the reciprocal ratio of the accelerations which the same motive force imparts to them (inert mass), and in the second place, as the ratio of the forces which act upon them in the same gravitational field (gravitational mass).'

In this passage the inert mass of a body is said to be the property the value of which determines the acceleration of the body under the influence of a given applied force. The gravitational mass, on the other hand, is said to be the property of the same body the value of which determines its weight in a gravitational field. The inert mass of a billiard ball, Einstein would have said, causes the ball to receive a finite acceleration when propelled by a billiard cue

123

over a smooth, level surface. In the absence of its inert mass the acceleration would be infinite. The gravitational mass causes the same ball to deflect a spring balance. In the absence of gravitational mass the ball would not be attracted by the earth. The distinction between these two kinds of mass is made even more precise in a later passage:

'Newton's equation of motion in a gravitational field, written out in full, is: *(Inert mass). (Accelerafion) = (Intensity of the gravitational field). (Gravitational mass).'*

The left hand side of the above equation defines the performance of a billiard ball when a force, F, is applied to it by a billiard cue. If the inert mass is m_1 and the acceleration f_1, one can write:

$$F = m_1 f \qquad \ldots \ldots \qquad (21a)$$

The right hand side defines the weight of the ball on a spring balance. Let this weight be W and the gravitational mass m_g The intensity of the gravitational field near the earth is g, so the algebraic expression for the right hand side is

$$W = g m_g \qquad \ldots \ldots \qquad (21b)$$

Equations (21a) and (21b) are found to give numerically identical results for the same ball when $f = g$. From this Einstein inferred that $m_g = m_i$. He spoke of 'the law of the equality between the inert and the gravitational mass' and made it clear that 'the numerical equality is reduced to an equality of the real nature of the two concepts'.

General relativity is based on this identity, so that Einstein could say:

'The possibility of explaining the numerical equality of inertia and gravitation by the unity of their nature gives to the general theory of relativity, according to my conviction, such superiority over the conceptions of classical mechanics, that all the difficulties encountered in development must be considered as small in comparison with this progress.'

This is very clear and, I think, irrefutable. We observe that the ratio m_g / m_i is always a constant and ask why. Einstein's answer is that, if general relativity is accepted, they are one and the same thing.

Zinzen, in the book referred to above, speaks, nevertheless, of 'a really great confusion of thought' about the nature of mass, and I have been led to the reluctant conclusion that he is right. For I have found very little appreciation of the distinction between attracting mass and the other two kinds. If the distinction is hardly ever mentioned, it might be because it is too obvious to need mention, and I thought at one time that it was so. In an earlier draft of this book I therefore took the distinction for granted, But comments from sundry authorities who saw the draft convinced me that the true reason why the distinction is not mentioned is that it is not even thought about. I was told, for instance, that the word 'mass' can never; have any other but the two meanings that figure in the above quotations from Einstein; that attracting mass and gravitational mass are known by everyone to be identical both numerically and in their nature; that this identity is proved by Newton's third law of motion, according to which action and reaction are equal and opposite; that general relativity does not rest on the identity of the first two kinds of mass only, but on that of all three kinds; that conceptual distinctions are hair-splitting and fruitless. All these statements are, I think, erroneous.

If the distinction between the three kinds of mass and its importance are clear to many, they are evidently not as universally known as they need to be. I have therefore no choice but to draw attention to some rather elementary facts about mass and gravitation before I can proceed further with the present inquiry. One of those unfortunate situations arises here that do occasionally arise in science when an obstacle to research and discussion is not raised by the difficulty of the subject but by an all-too-prevalent notion that the subject does not present any problem at all. Here the problem to be discussed is why the ratio of attracting mass to the other two kinds is observed to have a constant value. One cannot begin to answer this question until one has achieved an understanding of the nature of attracting mass, and in view of the widespread confusion that I have encountered about it I make no apology for presenting some elementary facts here.

Equation (21 a) represents the performance of a billiard ball in one kind of circumstance, namely, when it is being propelled by a cue. Equation (21b) represents its performance in another kind of circumstance, namely, when it is being weighed on a spring balance. In the first instance the performance is accelerated motion, in the second, deflection of the spring.

Einstein did not need to ask questions about the circumstances. He did not ask, for instance, how a push came to be given to the billiard cue. The question would be as irrelevant to his theme as the names of the players. Nor did he ask what was the source of the gravitational field in which the ball is weighed. It sufficed for his purpose to note that there was a field. He needed to speak of its intensity only, not of its source or cause. But it is now necessary for us to turn our attention to the source.

The intensity of the gravitational field has the letter symbol g. If M is the mass of the earth, one can write

$$g = GM / x^2$$

where G is the gravitational constant and x the distance between the centres of gravity of the earth and the ball. The question now is what suffix to give to M. The mass is the property of the earth by virtue of which it attracts the billiard ball. It is what Hamel called the 'attracting mass'. So I shall use suffix a. For the case when $f = g$ Newton's law can then be expressed algebraically in such a way that all the three kinds of mass are distinguished by their appropriate suffixes as follows:

$$F = m_i\, g = W = (\, G\, M_a\, /\, x^2\,)\, m_g \qquad ...... \qquad (21c)$$

21.2:   *The   Behaviour   of   Masses   in   Newtonian   Mechanics* Equation (21c) expresses the behaviour of masses in Newtonian mechanics when appropriate suffixes are used to distinguish between the three meanings of mass. Translated into words the equation yields three important statements in each of which the word mass has a different one of its three meanings. They are:

(1) The velocity of a body having inert mass is changed only when a force is applied to it. The relation between the force and the acceleration is as given by equation (21a), which shows that, for a given acceleration, the force is proportional to the inert mass, m_i of the body.

(2) A body having gravitational mass experiences a force when it is in a gravitational field. The force is proportional to the gravitational mass of the body, m_g , and the intensity of the field, as shown in equation (21b).

(3) A body having attracting mass is the origin of a gravitational field, the intensity of the field being proportional to the attracting mass, m_a.

Be it remarked in passing that Newton did not use the word 'field', but this word is today a suitable one for making statements in Newtonian mechanics.

Here is a definition of inertia: The property of a body by virtue of which it resists a change in its velocity. Hence statement (1) is true by definition. The measure used to express the magnitude of the inertia is, in Newtonian mechanics, the force needed to procure unit acceleration. The formula (21a) can thus be used as a means of defining this force. It expresses the definition in quantitative terms. Statement (1) is a means of recognizing inertia when one meets it and does not do more than define inert mass. To make the first statement is but a way of saying that inert mass is inert mass. Like many useful statements in physics it is a tautology.

If one defines gravitational mass as the property by virtue of which a body experiences a force in a gravitational field, the second statement is also a tautology. It is true by definition of gravitational mass. But it is not a tautology to say that inert and gravitational mass are always equal. In Newtonian mechanics this statement appears as a necessary clause in the Cosmic Statute Book. It was Einstein who showed that the equality need not appear there as it can be inferred from the more basic principle of general relativity. This was a great step towards the unification of physics.

The third statement is also a tautology in so far as it serves to define attracting mass. Again, what is not a tautology is the statement that there is a constant ratio between the attracting mass of a body and its other two masses. This is the fact that can at present only be gleaned from observation. It is not implicit in the definitions of any of the three kinds of mass. The definition of inert mass tells us that a body possessing inert mass will be accelerated if it is pushed or pulled by something. The definition does not tell us that the mass will itself do any pushing or pulling. The definition of gravitational mass tells us how a body possessing this kind of mass is *influenced* by its environment. The definition does not tell us how a body having gravitational mass *influences* its environment.

The distinction can be expressed in a different way. Inert and gravitational mass can be measured only by observing the body that possesses them and in order to make the measurement something has to be done to this body. It has to be pushed or pulled in order to measure the inert mass and it has to be placed in a gravitational field in order to measure the gravitational mass. Attracting mass on the other hand can be measured only by observing its effect on a different body and in order to make the measurement something has to be done to this other body. Thus the inert mass of a train could be measured by doing something to the train itself, namely by accelerating it and measuring the tension in the draw-bar. The gravitational mass of the train could be measured by placing it on a weighing device. But one could not measure the attracting mass of the train by any observation on the train itself. One could, however, measure this kind of mass, in theory though not in practice, by bringing a different body, a pendulum for instance, near the train and observing the effect of the train's attracting mass on the deflection of the pendulum. The gravitational mass is measured by the effect of the earth on the train, the attracting mass by the effect of the train on some other gravitational mass. (This other gravitational mass might be the earth, as will be shown later).

One could express the difference, if perhaps a little loosely, by saying that inert and gravitational mass are *properties* of the body while attracting mass is a *capacity* for affecting other bodies. Or one could say alternatively that inert and gravitational mass are observed while attracting mass can only be inferred. But how best to express the difference is not as important as to remember that there is a difference. Use of the one word 'mass' for things that are conceptually so distinct has had some unfortunate results.

21.3: *Mutuality in Gravitation*

The notion that the equality of action and reaction proves the identity of gravitational and attracting mass is based on a misapprehension about elementary mechanics. But as it seems to be rather prevalent it ought not to be ignored.

Let us return to equation (21c). It gives a numerically correct result; so it contains everything needed for weighing the billiard ball. But it fails to embody a statement that is to be found in elementary textbooks. This is that the billiard ball attracts the earth at the same time as the earth attracts the billiard ball. This statement attributes an attracting mass, m_a to the ball, which does not appear in the equation. We must find out where and how to show it.

Mutuality implies that the ball attracts other objects with a faint gravitational field. So we can represent the field intensity of the ball as g_m

We then have to write g_M for the field intensity due to the earth. The intensity, g_m is of course very small indeed and can be neglected for practical purposes. But when the ball is replaced by the moon it is no longer so. The moon does not have an orbit around the centre of the earth but around the common centre of gravity of both bodies, and this is because the moon has a significant field of its own. In computing their relative movement one gives an acceleration, g_M to the moon toward the earth and another acceleration, g_m of the earth towards the moon The case of the billiard ball is similar. When the billiard ball is free to move, it does so in the earth's field, g_M, while the earth moves towards the ball in the field of the latter, g_m. If one wants to be pedantically accurate one must therefore define the relative motion of the billiard ball and the earth as determined by

$$g = g_M + g_m \quad \text{......} \quad (21d)$$

In this equation

$$g_M = k\ M_a\ /\ x^2$$

$$g_m = k\ m_a\ /\ x^2$$

The pull of the earth on the ball is g_M, m_g , while the pull of the ball on the earth is g_m, M_g. The very small field intensity due to the ball is compensated for by the large attracted mass. One can say that the heavy earth is being weighed in the faint field due to the ball at the same time when the light ball is being weighed in the strong field due to the earth. In these circumstances both show the same weight and contribute equally to the deflection of the spring. The ball presses downwards on the top end of the spring and the earth presses upwards on the bottom end. In producing a compression the forces are additive; hence the plus sign in equation (21d).

Let the pull of the Earth on the billiard ball be F_e and the pull of the billiard ball on the Earth F_b. One can then write

$$F_e = (kM_a/x^2)m_g$$
$$F_b = (km_a/x^2)\,M_g$$

These two forces are equal and opposite. Action and reaction are also equal and opposite. Therefore it has been argued, to my surprise, these forces are action and reaction. But are we, according to this argument, to regard F_e as action and F_b as reaction, or vice versa? The difficulty of saying which is which ought to serve as a warning against so careless a conclusion.

The true situation can be understood with the help of the diagram in Fig. 4. F_e is shown as acting on the top of the spring and has its own equal and opposite reaction. The same holds for F_b, which is shown as acting on the bottom of the spring. Each force could equally well be shown in any other place between the ball and the Earth. It is only for convenience of presentation that they are shown separated. The important fact is that if either F_e or F_b disappeared, the other force would still be there with its own reaction.

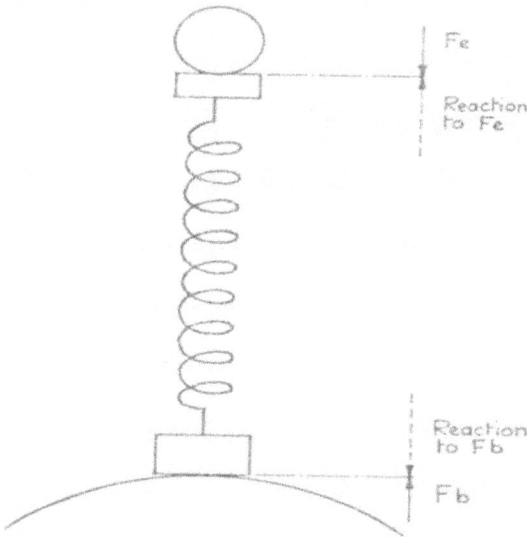

Fig. 4. Action and Reaction produced by gravitation

The reader may have difficulty in believing that the very weak field of the billiard ball has any effect at all on the spring. It will help him if he imagines that mass is being continually transferred from below the spring balance to the tray where the billiard ball rests. If this goes on until half the mass of the Earth is on the tray and half left below the spring balance, the situation will be reached when half the Earth is weighed in the field of the other half. A gramme mass will then have half a gramme weight. If the process continues until no more than a billiard ball is left beneath the spring balance, practically the whole of the Earth will be weighed in the field of this billiard ball. In this field, a gramme mass will weigh very little. The compression of the spring will be the same as before the transfer of substance was begun.

From the above considerations it follows that the pedantically precise algebraic expression for the weight of the billiard ball is

$$W = (1/2GM_a/x^2)m_g + (1/2Gm_a/x^2)M_g \dots (21e)$$

from which it appears that k, above, equals $1/2G$.

Equations (21e) and (21c) are numerically equal, but this is only because $M_a\,m_g = m_a\,M_g$. One cannot infer from this equality that attracting and gravitational mass are of the same nature.

Their numerical equality, it must also be noted, has been obtained by a suitable choice of G. If one had arbitrarily chosen G as unity, one would say that one unit of gravitational mass is always equal to G units of attracting mass.

The value of G has been determined by experiments with more or less massive spheres by Cavendish, Poynting, Boys, and others. These experiments show that the spheres have attracting mass as well as inert and gravitational mass. But they do not prove that less massive bodies also have attracting mass. It is here that equation (21e) is valuable. If one were weigh a body that had only gravitational mass and no attracting mass, it would have half the observed weight. But the smallest objects that give observable deflection on a spring balance have the weight that one would predict on the assumption that they have attracting mass. One may therefore safely conclude from observation and experiment that the law of constant proportionality between inert, gravitational, and attracting mass holds for very small accumulations of matter.

It is, nevertheless, important to bear in mind that this mutuality in gravitation is based on observation and experiment, and not derived from any more fundamental principle — at least in Newtonian mechanics. If one were to find a body that had no attracting mass, it would not be weightless. Action and reaction would apply to its pressure on a spring balance. Every other principle of which I am aware would also be preserved. I can think of no way of proving the impossibility of a body that has no attracting mass while having gravitational mass except by the inductive kind of reasoning that takes the line: This has never been observed; therefore it cannot happen.

Before I leave the discussion of the three kinds of mass in terms of Newtonian mechanics, I should like to point out that one cannot discover whether a body does or does not possess attracting mass by observing it while it falls freely. One can do so only by weighing it. For the equation of the free fall is

$$gm_i = (GM_a/x^2)m_g \dots \dots . \text{(21f)}$$

The attracting mass, m_a, of the falling body occurs here only in so far as it influences $g = g_M + g_m$ to a minute extent. A ball that had no attracting mass would fall with practically the same acceleration as an actual ball. One could observe the lack of attracting mass in a falling body only if the ratio g_m/g_M were large enough to be significant.

It should be remembered that it was from equation (21f) that Einstein developed general relativity. This equation established for him the identity in nature as well as numerically, of the inert and gravitational mass of mass m, namely m_i and m_g. This equation gives no information about the attracting mass M_a. It is therefore impossible to reach any conclusion about the attracting mass of m so long as one uses equation (21f).

The question arises what part, if any, attracting mass plays in general relativity. Only experts can give the answer, but I should like to say that I have been able to find but little discussion of this question. This may, however, be due to my meagre reading. General relativity does explain quite clearly why a gravitational mass is accelerated when it finds itself in a gravitational field but, so far as I can make out, it does not account for (and does not need to account for) the field. What is basic in general relativity seems to me to be equally true whatever the object may be that is the source and cause of the field, by whatever process the field is produced.

Some relativists to whom I have spoken about this have taken it for granted that in all relativity equations, the symbol that represents gravitational mass automatically represents attracting mass as well. It is an assumption that calls for great caution. I am sure that the basic conclusion of general relativity does not depend on it. I do not know whether any subsidiary conclusions do. If so, I venture to suggest that they ought to be carefully scrutinized.

That general relativity does not depend basically on the identity of attracting and gravitational mass seems to me not only to follow from the equations but also to be apparent when one translates into the language of general relativity the three statements about mass that have been expressed above in the language of Newtonian mechanics. This will be done in a moment, but it is necessary to lead up to the translation by first explaining the nature of the gravitational field as it appears in general relativity.

21.4: *Gravitation as Interpreted in Relativity Theory*

In Newtonian mechanics, a gravitational field appears as a region in which force acts at a distance from its source. Action at a distance had always been regarded as an unsatisfactory concept. One thinks of a force as applied by physical contact between something and ponderable matter, such as when a thing is pushed by a stick or pulled by a string. The notion could never be easily accepted that a force could be exerted at a distance from any physical object. Yet this is what seemed to happen to a stone that was attracted by the Earth. To make this notion a little less unsatisfactory, an hypothetical ether was postulated. It was assumed that a gravitational mass produced some kind of effect, called a strain, in the ether and that the force was applied by physical contact between the stone and this strained ether. To give a more concrete meaning to this assumption, the strained ether was later said to contain tubes of force that extended between the attracting Earth and the attracted stone. Thus every massive body was believed to have a twofold environment: to be surrounded at the same time by a featureless, undifferentiated space and by a featured ether.

It has been said that Einstein abolished the notion of a featured ether. One might equally well say that he abolished the notion of a featureless space. He retained the concept of a featured environment with physical properties but showed that there was no need to assume an additional featureless environment. For this single environment with physical properties, he wisely retained the name space. But what he did abolish was the concept of space as the *container* of the material universe. Einstein's featured space is a *constituent* of it.

One of Einstein's great achievements was to define precisely the physical condition of a gravitational field. It had previously been vaguely thought of as a strain. He showed that this condition can be described as a departure from Euclidean geometry. Space that shows such a departure is technically called curved. Einstein further proved that the kind of motion that would appear to be at constant velocity in flat space would appear to be accelerated motion in curved space. If a stone moved out of nearly flat space into a gravitational field, as would happen if it fell from a very great height, its velocity would change, but that would not mean that it was subjected to a force. In other words, one must not think of the acceleration of a falling stone as the consequence of what is done to the stone but as the consequence of where it is, i.e. the acceleration is a function of the local geometry.

The new outlook made it necessary to define inertia more precisely than had been necessary with Newtonian mechanics. In relativity language, inertia is the property by virtue of which the acceleration of a body that is free of all restraint depends on the curvature of the space in which it finds itself. If the space has zero curvature, the acceleration is zero. In this respect, relativity is not a denial, but an extension, of Newtonian mechanics.

One can conceive a line in four-dimensional space-time that a freely moving body would trace when not subjected to a force. According to Newtonian mechanics, it would always be a straight line and represent a constant velocity. According to relativity theory, it would be a straight line only when the space was flat. When the space was curved, though the body was not subjected to a force, the line would be curved and represent an acceleration. Thus to the question: How can a falling stone be subjected to a force without physical contact with ponderable matter? Einstein's answer was as beautifully simple as it was unexpected. He said that the falling stone is *not* subjected to a force. A force is no more needed to maintain an acceleration in curved space than to maintain constant velocity in flat space. On the contrary, a force would be needed to prevent acceleration in curved space. A stone is subjected to such a restraining force when it is at rest on a shelf, the force being exerted by physical contact with ponderable matter, namely the shelf. In flat space, an observer who is being accelerated experiences a force. In curved space it would be an observer who was *not* being accelerated in the direction and at the rate defined by the curvature who would experience a force.

21.5: *The Behaviour of Masses in Relativity Theory*
Now let the three statements about mass that were formulated above in the language of Newtonian mechanics be translated into the language of relativity: 'The acceleration of a body having inert mass and not subjected to any force depends on the geometry of the space in which the body is. The acceleration is zero when the space is flat and has a finite value when the space is curved.' It would be idle to discuss whether this is a tautology. Whether it is or not, it is a definition of inert mass in the language of relativity just as the equivalent statement given earlier here is a definition formulated in the language of Newtonian mechanics.

'A body that is free of restraint is accelerated when it is in the curved space that constitutes a gravitational field.' This is but to mention a special case of (1).

'A body having attracting mass is itself the cause of a curvature in space, the value of the curvature being proportional to the value of the mass.' This is no tautology. It is not implicit in the definition of inertia. There does not appear to be any reason why a body that experiences an acceleration when the space around it is curved should itself have any influence on the curvature of the space. No one has been able to show that the one fact is a logical consequence of the other. Though relativity has explained much, it has not explained the apparently invariable association of inert and attracting mass. One is still left with the question: Why has one never observed an inert mass that leaves the space around it undisturbed?

It has become clear that both Newtonian mechanics and general relativity leave us with the same puzzle about attracting mass. Observationally, this and the other two kinds of mass, inert and gravitational, are always coupled. But no reason has been given why it should be so. Why, one is led to ask, does one never observe a body that has inert and gravitational mass but has no attracting mass? Why, in relativity language, does one never observe a body with inert mass that leaves the space around it undisturbed?

I shall suggest a possible answer later on, but much ground remains to be covered before this can be done.

21.6: *Two Observers or One?*
After this book had been sent to the printer, I came by accident upon two papers published in the *Physikalische Zeitschrift* of 1921. They are by one E. Reichenbacher. In these papers, much is said that has been said in the present chapter. It is said very clearly and cogently, if in a different way. Reichenbacher urges many sound arguments against identifying the gravitational with what he calls, rather unfortunately, the inert field. At the same time, he shows complete understanding of the reasoning that led Einstein to identify inert with gravitational mass. Reichenbacher distinguishes between passive and active gravitational mass. This is the distinction that has been made in the present chapter between gravitational and attracting mass.
It is strange indeed that this distinction has not received greater prominence in spite of the fact that attention was drawn to it with some insistence in the very early days of general relativity. In these two articles, Reichenbacher is revealed as a clear and powerful thinker. To understand why his contribution was so ineffective, one must recall the atmosphere in the early nineteen-twenties.

Concerning general relativity theory, scientists found themselves in two opposing camps. Those in one camp rejected the whole of the theory, and not without emotion. Those who did not express themselves violently on the subject did so peevishly. Scientists in the other camp supported relativity theory equally vehemently. Aware of the immense clarification and unification that it had achieved, they saw themselves as crusaders for a new revelation. Loyalty to the cause, loyalty to Einstein, impelled them. Those who did not in the least understand relativity theory rejected the whole of it. Those who did more or less understand it accepted it in its entirety, often uncritically. As defenders of the Faith, they could tolerate no doubts. They reacted, as we all do at times, against the nagging suspicion that something needed further clarification.

In such an atmosphere, Reichenbacher could hardly escape being unpopular with both camps. He understood general relativity so well that he could venture to explore its limitations. For him, doubts about the identity of the inertial and the gravitational field were not nagging. He faced them squarely. In doing so, he came to realize that the gravitational field can be described as a distortion of space such as to give it a non-Euclidean geometry, and that this distortion is independent of an observer in a way that causes the gravitational field to differ from the 'field' that he saw as the result of accelerated motion and that he called one of acceleration. That Reichenbacher's analysis led him to a clearer explanation of centrifugal force in relativistic terms than has been provided by other relativists makes it all the more regrettable that his work should have received so little notice. Relativists of his time would seem to have disliked the awkward questions that he propounded all too inescapably. Unable to refute him they ignored him.

I am bold enough to suggest that this reveals one of the less healthy aspects of the science of relativity. Today, opportunities for sanitation are rarer than they were in the nineteen twenties. In those years, many of the best intellects trained in physics devoted a great deal of their time and thought to relativity theory. Today, experts in this field are but few. Special relativity occupies no more than a minute fraction of the physics syllabus in our universities, and general relativity is hardly taught there at all. This is unfortunate. We are in danger of losing the great benefit, the considerable insight, that relativity theory brought. It has much more to tell us than the formulae expressing the Lorentz transformation. I am hoping that the following chapters will illustrate how a revival of interest in general relativity can prove rewarding.

If different interpretations of general relativity were possible in 1921, they are certainly still possible today: for there is little, if anything, to show that the difficulties to which Reichenbacher referred have been resolved — or even much appreciated. There can be no change unless interest in general relativity becomes more widespread, and it may encourage some to turn their attention to this subject if I give a very brief account of its basis. For this purpose, the mathematical treatment is not relevant and I shall not attempt it. I shall follow as closely as possible the lines taken by Einstein and other brilliant expositors of the past, while departing from these lines just enough to show why those who ignored, and have continued to ignore, the difficulties have been guided not by understanding but by careless thinking.

In the early days of relativity, we were introduced to the notion of a box in space, far distant from any ponderable matter. Einstein was the first to invent this illustration. A man was supposed to be inside the box and to make all the observations that could enable him to learn something about its motion through space. The man could secure himself to the floor of the box. He could climb to the ceiling and catch hold of a strap attached to it. He could let go of the strap and observe what happened. He could measure any forces that were exercised on his body. When he had observed and measured, he could apply logic and mathematical reasoning to the results and arrive therefrom at some conclusions about the nature of space, time, and gravitation.

It will be convenient to refer to the observing, reasoning man in the box by a name. I should call him Mr.. Einstein were it not that I might then mislead the reader and also divert attention from the subject of relativity to the question what, in fact, Einstein did or did not say, what he did or did not think, what he ought or ought not to have said or thought. For what I have read and what I have learnt in conversation has shown that different people have different ideas about these matters. So the man inside the box shall be called Mr.. Smith.

Let us begin by considering an occasion when the box is in uniform motion. Mr. Smith tries to measure its velocity. He would like to select a point in space outside the box and to measure its velocity relative to that point. But he finds, to his disappointment, that he cannot even define a point in space in any helpful way, let alone measure a velocity relative to it.

One point in empty space is exactly like any other point. It can be defined only as its distance from a reference object and the only available reference object is the box. Mr. Smith could choose a point defined as being at a particular distance from this. But then the point remains, by definition, always at the same distance from the box, whatever the speed of the box may be. If the box is considered to be moving, so is the point. If the box is at rest, so is the point.

Mr. Smith cannot avoid the conclusion that he has no means of knowing whether his box is at rest or in uniform motion. It is, indeed, meaningless to ask the question. In the technical language of relativity, there is no privileged frame of reference by which to define a uniform velocity. Mr. Smith is perfectly free to ascribe to the box any velocity that he likes. The figure that he mentions cannot be disputed.

Relativists have another way of expressing this. Let us suppose that someone has constructed a four-dimensional model of the space-time in which the box is situated. One cannot, of course, make such a model out of wood or metal; but one can represent it by algebraic symbols. For our purpose, such a representation is just as serviceable. Nothing would be gained here by writing the symbols down. It suffices that it can be done and that the space-time relations can be expressed mathematically as clearly and precisely as they would be if a material model could replace the algebra. How this is done is not relevant.

If successive positions of the box in uniform motion are represented by points in the model, the points lie along a straight line. This may be vertical or sloping; it depends on how the model is tilted. One can, if one likes, adopt the convention that zero velocity is represented by points that lie vertically above each other. Points along a sloping line would then represent a finite velocity, which would be greater the more the line departed from the vertical.

If Mr. Smith knew which was the right side up for his model, he could then read the velocity of the box off from it by measuring the slope of the line. But he finds that there is no means of determining a right side up. He is free to tilt the model in any way he likes. He can so tilt it that the line representing successive points of the box is vertical and say that it shows the box to be at rest. But he can also tilt it so as to show the box to be moving at a uniform velocity. No inclination of the model could be claimed to represent a wrong velocity, which is another way of saying that there is no such thing as absolute velocity.

This is the basic tenet of special relativity, but not, of course, the whole of it. What has been said above does not show the part played in special relativity by the velocity of light. But this omission need not detain us. Our immediate concern is with general relativity.

Let us next suppose that there is a hook in the ceiling of the box on the outside and that a rope is attached to this hook. Einstein asked us to imagine that the rope is being pulled, and pointed out that it is not relevant to say by what agency. For descriptive purposes it is, however convenient to name the pulling agent, so I shall say that an angel is pulling the rope from time to time and with a constant force. As a consequence the box, with Mr. Smith inside it, receives at intervals a constant acceleration.

Action and reaction being equal and opposite, Mr. Smith is subjected to a force in the opposite direction from that in which the angel is pulling. While the pull is upwards away from the ceiling the force observed by Mr. Smith is downwards towards the floor. Being inside the box, Mr. Smith cannot see the rope and does not know what causes the force on his box He reasons as follows:

'A moment ago there was no force, now there is one. So there is a physical change to the system in which I find myself. What is the cause of this change? I can think of two possible ones. The first is that this box is being accelerated in the direction towards the ceiling. The second is that some undetectable device is attached to me and is pulling me downwards towards the floor. Have I any means of discovering which of these two theories is the correct one?

'Can my model of space-time help? If the first theory is the true one, if my box is being accelerated, successive positions when plotted in the model will not lie on a straight line. They will lie on a parabola. Let me do the plotting. Now let me tilt the model in such a way that it represents my box at rest. I find that it cannot be done. I can tilt the model so that a little bit of the line connecting the points is vertical. But then all the remainder of the line is sloping. Velocity can be made to disappear on the model, but not acceleration. The inference is that a body in uniform motion can rightly be regarded as at rest all the time;

but a body in accelerated motion can be regarded, at most, as at rest for a moment of time.'

It now occurs to Mr. Smith to construct a different model of space-time. He calls it model B and calls the first one model A. In model A, equal distances in space-time were represented by equal distances in the model. In model B, it is not so. Distances are represented to functional scales, somewhat like log scales on graph paper. The selected scales are unusual and represent relations in non-Euclidean geometry. They are arranged in such a way that successive positions of the box lie along a curved line when the box is in uniform motion and along a straight line when the box is being accelerated at a uniform rate. As for model A, there is no reason why model B should be tilted in one way rather than in another. Like the first model, the second one correctly represents the fact that there is no absolute velocity.

The second model can be so tilted that the points lie along a vertical straight line. It then represents a condition when the box is at rest. Let us learn how Mr.. Smith reasons about this:

'When I constructed my second model I thought of it only as a convenient plotting device that would enable me to place all the points along a straight line. I regarded model A as the true scale model and model B as one constructed deliberately to be out of scale, to be a distorted representation of reality. But can I have been wrong in this? Is model B more than a mere plotting device? Can it be a true-to-scale model of the space-time in which my box is situated? Can space-time have the strange shape that is represented by the model? If so, space-time itself has a physical property, namely curvature.

'This suggests a third way of accounting for the force that I am experiencing. The reason may be neither that the box is being accelerated upwards nor that I am being pulled downwards. That notion that I am being pulled is unsatisfactory anyhow, so long as I cannot detect any devices attached to me by means of which the pull can be transmitted to my body. So the third explanation, which has only just occurred to me, deserves consideration. It is that the box is in a region where the space-time is non-Euclidean.

'If this is the explanation, the box is not being accelerated. It can be regarded as either at rest or in uniform motion. Model B correctly represents the situation. I can tilt this model so that the line is vertical if I like. So I shall give up the old notion that something is pulling me downwards. It ought to be discredited anyhow. I shall consider instead only the two possibilities that remain, namely that the box is being accelerated in a space-time correctly represented by model A or is at rest in a space-time correctly represented by model B.

'Can I make any observation here and now by means of which I can decide between these alternative possibilities?

'No. The force exerted on my body must be just the same whether I am being accelerated in flat space or am at rest in curved space. The only difference that I can mention is that I attribute the force to the inert mass of my body when its cause is an acceleration, and I attribute it to a different property (I shall call it the gravitational mass of my body) when the cause is the curvature of space.

But if the only difference between two properties is the name that convention has given them, there is no real difference. So it is reasonable to regard inert and gravitational mass as identical.'

By gaining this piece of insight, Einstein took a great step forward towards the unification of science. What has been said here so far conforms to the traditional view of general relativity and differs only slightly from the conventional manner of its presentation. But it now becomes necessary to consider some of the false conclusions at which one can arrive if one thinks carelessly and superficially. If any of those who speak on relativity theory with authority have in fact arrived at any of these false conclusions, as I fear may have happened, it has become very necessary to do some rethinking. If I am wrong and no errors have been made, it will nevertheless help towards clarity to appreciate what errors could be made.

Mr. Smith might be tempted for a moment to think that it was meaningful to speak of a 'field of acceleration' in the sense in which one speaks of a gravitational field. (After all, Reichenbacher did so.) Mr. Smith would suggest therewith that a field of acceleration occurs in flat space when a force is the result of an acceleration just as a gravitational field is present in curved space. But he would, I hope, soon realize that such a manner of speaking was misleading and imprecise. The word 'field' should be retained for the *condition* of a region and not for what happens in it. The acceleration of a body does not change anything in its surroundings. If the space is flat, it remains flat when something in it is being accelerated.

Should Mr. Smith persist however, in speaking of a field of acceleration, he might fall into the further error of thinking that the space around him really did change its curvature when a body was accelerated in that space. Suppose he were to climb to the ceiling of his box and hang there from the strap for a short while. If he let go, his body would be accelerated relative to the box. It would be so whether the box was being accelerated in smooth space or was at rest in curved space. If it was the former, Mr. Smith would be at rest after he had let go while the box would continue to be accelerated. If it were the latter, Mr. Smith would be accelerated after he had let go, while the box would continue to be at rest. By falling, Mr. Smith does not smoothe out the space that surrounds him. He just falls.

The error of thinking that the act of falling smoothes out space-time can arise only if one believes that the choice between model A and model B depends on the motion of the observer. But it will have become clear that it does not. As I have said already, Mr. Smith can find nothing inside his box that will enable him to know which model to choose. His own motion relative to the box does not tell him; nor do any other observations made there.

Having appreciated the fact that he cannot alter the curvature of space-time

139

around him by falling, any more than he can by dropping his watch, Mr. Smith will replace his previous assertion by the following:

'When I feel a force on my body and still have hold of the strap, my successive positions lie along a curved line if plotted in model A and along a straight line if plotted in model B. After I have let go I do not feel any force and the shape of the lines in the model is the other way about. While I am falling my successive positions are represented by a straight line in model A and by a curved line in model B. I can represent this conveniently in tabular form as follows:

Circumstance	Condition of space-time	
	Smooth	Curved
Correct model	If A	If B
Body subjected to a force when	Accelerated	At rest
Body not subjected to a force when	At rest	Accelerated
Uniform motion represented by	Straight line	Curved line
Accelerated motion represented by	Curved line	Straight line
Force attributed to	Inert mass	Gravitational mass

'I must refrain in future from saying, or implying, that my actions change the curvature of space-time. As I do not even know from any observation that I can make from where I am, whether the correct model is A or B, I can certainly not claim to know that when I am falling I am changing from the one model to the other.'

There is the possibility of yet another error. Mr. Smith may reach the erroneous conclusion that he possesses no means at all of deciding between the two models. He may, in other words, believe that observations made by an observer outside the box are as inconclusive as those made by himself.
If he makes this mistake he will feel justified in asserting that the two models, as any others, are equally valid in the same way as two different tilts of either model are equally valid. He will then enunciate the theory of full equivalence between a body at rest in curved space-time and a body in accelerated motion in smooth space-time. But Mr. Smith is fortunate in having a friend outside the box from whom he can obtain some useful information.
The friend's name is Mr. Jones. He is standing beside the angel. The two gentlemen establish telephonic communication. The following conversation takes place:

SMITH: 'What is happening where you are?"

JONES: 'The angel is pulling himself hand over hand along the rope towards your box.

SMITH : I believe you are wrong. It seems to me that the angel must be pulling the box towards himself

140

JONES : 'Have it that way if you like. All that I can observe is that the rope is taut and is getting shorter. It is doing so at a rate of 9.80665 metres per second per second. The explanation might equally well be that the angel is being accelerated towards the box or that the box is being accelerated towards the angel. I have no means here of knowing which it is. The two statements are fully equivalent. To ask which it is is a meaningless question.'

SMITH: 'But I have means of knowing which is which. I am experiencing a force towards the floor of the box. As my body has inert mass I conclude that it is the box and not the angel that is being accelerated.'

JONES : 'After what you have told me, that seems to be the more probable explanation. My observation alone left the question open, but our joint observations make it reasonable to assume that the box is being accelerated and not the angel.'

SMITH: 'Yes. And my observation alone left the possibility open that there was no pull at all on the rope but that, instead, the box was in a non-Euclidean region of space-time. Our combined observations have provided the answer to that question too.'

At this moment it occurs to the two observers that Mr. Jones would have known whether the angel was being accelerated or not if the angel had possessed inert mass. They agree that acceleration of a body P relative to Q is not equivalent to acceleration of Q relative to P. If the bodies are in smooth space-time, the force on each is the product of its acceleration and its inert mass. If the acceleration is zero, its product with inert mass is also zero. Conversely, if the force is zero and the space-time is smooth, the acceleration is also zero.

After a while, the angel gets tired of intermittently pulling at the rope. He changes his position and places himself underneath the box. Mr. Jones accompanies him. Presently Mr. Smith calls Mr. Jones on their telephone.

SMITH: 'Jones. The angel is pulling at that rope again.'

JONES : 'No he is not. The rope is curled up neatly and lies on top of the box. The angel is underneath the box. I am here with him.'

SMITH: 'What has the angel been doing?'

JONES : 'He has placed a big sphere underneath the box. Its circumference is 40,000 kilometres.'

SMITH: Is that all?'

JONES: 'Yes. All I can see.'

SMITH : 'As the angel is no longer pulling at the rope, I must assume that the box is no longer being accelerated. And yet I feel that same force as I did while the acceleration lasted. So I have to conclude either that some undetected device is attached to my body and pulling downwards or that the box is in a non-Euclidean region of space-time. I deprecate the mystical hypothesis of undetectable devices and so I have to assume the latter. Something must therefore have changed the physical nature of space-time around here in such a way as to cause curvature. The only physical change that you can report is the appearance of the big sphere. So I conclude that this is influencing the geometry of space-time in its vicinity.'

Mr. Smith and Mr. Jones confer further together and reach agreement on several points. They are satisfied that there is no difference between inert and gravitational mass; that there is nevertheless a difference between acceleration and uniform motion; that relative acceleration has real meaning; that the process that causes a stone to be subjected to a force during its acceleration is not the same as the process that causes it to be subjected to a force while it is at rest on a shelf; that an observer may be conceived to be so situated that he cannot discover which of the two processes is operating; that an observer may also be so situated that he can discover just this; that, for instance, Smith inside the box can only observe the effects of the processes, which are identical for both, while Jones outside the box can observe the sources of the processes, which are different for both; that it is the big sphere underneath the box that causes the geometry of space-time to become non-Euclidean.

Having thus clarified the subject they are able to formulate the big question that had previously been obscured by sundry misconceptions:

By what process is the distortion of space-time effected?

Chapter 22

Unsolved Problems of Gravitation

22.1: *Gravitation Appears to be an Isolated Phenomenon*
One of the major problems presented by gravitation has been discussed in the preceding Chapter 21. But there are others and these will now be examined.

One of them is how to bring gravitation into association with other natural phenomena. At present it appears curiously isolated from them. In this it is unique. Mention has already been made in Chapter 1 of the unification of the various branches of physical science that has been achieved during the past few centuries. Heat, light and sound, electricity and magnetism, mechanics, chemistry are now all under one common roof. Quite a lot is known about the relation between them. One can often express observations made in one of these branches in terms appropriate to another one. Chemical processes are explained with reference to electrical forces. Light is interpreted as an electromagnetic phenomenon. The effect of a magnetic flux on the polarization of light is a subject for study. The temperature of a gas is attributed to the momentum of the constituent molecules, a mechanical effect. Wherever one turns in physical science one meets this same close interlocking between its various branches. More and more generalizations in physics are, let me repeat here, found to be logical inferences from other, more basic, ones.

Gravitation is one of the few phenomena in physical science that have entirely escaped this process of unification. It does not interlock with any branch. It stands outside the roof that covers all the others. The few generalizations that can be made about gravitation have not been found to be logical inferences from any more basic generalizations. Gravitation is, in a word, still one of the phenomena that cannot be explained.

I propose to show in this Part Four that it need not continue to be so. Gravitation, like so many other natural phenomena, will be shown to be a logical inference from the Principle of Minimum Assumption.

22.2: *The Principle by which a Process Tends to Reduce its Cause*
A third puzzling fact about gravitation is that it seems to violate one of the most basic principles known to physics. This is that in a self-contained system every process tends to be such as to reduce its cause. Something has already been said about this principle in Chapter 3, when the reasons for rejecting

hypothesis (A1) were presented. It will be convenient to repeat a little of this here.

The movement of electric charges in an electrostatic field provides an example of a process that reduces its cause. At the origin of the field is an accumulation of electrons. As like charges repel each other, any of those that are free to move are propelled out of the field, thus weakening it. Any unlike charges, on the other hand, that come into the field are attracted. They neutralize the charge at the origin, again with the consequence that the field is weakened. In this example the process is the movement of electric charges; the cause is an electrostatic field; and the process reduces the intensity of this field.

Another example is the transfer of heat between surfaces. The cause is the difference in the temperature of the surfaces. As the process of transfer continues, the temperature difference decreases and the rate of heat transfer becomes slower.

The principle can be further illustrated. Most chemical substances are reactionable, which means that they enter into combination more or less vigorously with other substances. But when they do so the resultant products are always less reactionable; the process of chemical reaction tends to reduce the capacity for reaction. Oxygen and iron, for instance, combine with some vigour at certain temperatures, but rust, the product of their combination, is comparatively inert.

Causes of physical change can be expressed in terms of gradients. When an electric charge moves, it follows an electrical potential gradient. When heat flows from one place to another it follows a temperature gradient. In chemical reactions there are strong potential gradients between atoms and molecules. According to the principle being discussed here, we live in a world in which gradients are flattened out as time goes on.

If the opposite principle prevailed all causes would become more and more pronounced in the course of time, gradients would become ever steeper and steeper; the world would become a place of ever-increasing contrasts. Every electrostatic field that had begun as a weak one would have been becoming stronger and stronger in the course of time without limit. Every hot surface would be getting hotter as it radiated heat and every cold surface would be getting colder as it received it. We should be living in a universe in which chemical processes were becoming ever more violent. Such a universe would be unstable and so the principle by which processes reduce their causes can be called the Principle of Stabilization. As there will be occasion to refer to it again, this name will be useful.

This principle is, of course, only observable so long as the system is self-contained. If, for instance, an electric generator supplies fresh electrons as fast

144

the old ones are removed or neutralized, there is no reduction in the intensity of the electrostatic field.

The Principle of Stabilization is generally regarded as even more basic than the first and second laws of thermodynamics, for it embodies both. If starting a process sufficed to increase its cause, perpetual motion would be possible and the first law would be refuted. If the transfer of heat caused temperature differences to augment, every thermal change would be accompanied by a decrease in the entropy of the system and the second law would be refuted. To deny the Principle of Stabilization would be heresy indeed.

And yet! Gravitation has the appearance of violating it. When bodies are free to move in a gravitational field they fall on to the attracting mass; but they do not thereby reduce the intensity of the field, as electric charges do when they fall in an electrostatic field. The bodies that fall in a gravitational field add to the mass at its centre and thereby increase its intensity. We live in a world in which all gradients tend to become flatter in the course of time with the exception of gravitational ones. These seem to tend to become steeper.

The implications of this strange characteristic of gravitation cannot be dismissed lightly. Every accumulation of matter must be attracting neighbouring bodies towards it and, in doing so, becoming increasingly powerful. It will be shown in Appendix B that, as a consequence, the cosmological model that is based on Asymmetrical Impermanence does not resemble actuality. In this model the whole of the matter in the universe would eventually fall together into one single unit. The reason is that any two neighbouring accumulations of matter may be regarded as in competition for the substance that occurs between them. The larger accumulation will always acquire a greater proportion of it and thereby become able to compete still more successfully with its smaller neighbour. In the absence of anything to counteract this tendency for the larger concentration always to grow at the expense of the smaller ones, the largest of all must eventually swallow the others one by one.

It would be easy to show that the models based on either hypothesis (A1) or hypothesis (A2), as listed at the beginning of Chapter 3, would be similar in this respect. Only the model based on the combination of (A3) with (B3), on Symmetrical Impermanence, will, if the reasoning in Appendix B is confirmed, continue to contain discrete concentrations at all times. But the reason is, it will be found, that every large concentration, every galaxy, finds itself at intervals of about three-and-a-half thousand million years in a position where the loss rate from extinctions within its domain exceeds the gain rate from new origins. The galaxy can, in consequence, capture less new matter than it loses and so must dwindle for a while.

If one rejects Symmetrical Impermanence, one can only arrive at a model that

145

resembles actuality by inventing sundry *ad hoc* hypotheses. Perhaps, one has to speculate, the expansion of space is effectively opposing gravitational forces, a surmise that cannot easily be justified quantitatively.

Or perhaps some other, as yet unknown, force operates in extragalactic space to keep the galaxies apart. Or perhaps such a force operated once-upon-a-time only in the remote past and caused all celestial bodies to acquire a momentum away from each other. If quantitative reasoning leads to an unwelcome conclusion one can always, as I have said repeatedly in these pages, save the favoured view with the help of sufficient hypotheses.

But those who adopt these, as also those who prefer to base their cosmological model on the Principle of Minimum Assumption, must all feel some uneasiness at the apparent failure of gravitation to conform to so basic a principle as that of stabilization. A single exception to a principle that is otherwise absolutely universal calls for inquiry.

22.3. *Gravitational Fields are Unidirectional*
Gravitational fields always have the same sign. In this they differ from other known fields. In electrostatics like charges repel and unlike ones attract each other. Hence in some fields a given particle moves towards the centre of the field and in others the same particle moves away from the centre. This is expressed by saying that electric fields may be either positive or negative. It is the same for magnetic fields. A north-seeking pole moves in one direction in one kind of field and in the opposite direction in another kind. But it is not the same with gravitation. Like gravitational masses do not repel but attract each other, and unlike masses are not known. In other words, electrostatic and magnetic fields are described sometimes as positive and sometimes as negative, but one has no reason to describe gravitational fields in similar terms. These are observed to be unidirectional; bodies move always towards, and never away from, their centres.

This does not appear to be the logical consequence of any other known fact. It cannot be deduced from the definition of gravitation; it is not, so far as can be ascertained, a tautology. No reason has been found why inert masses should not do what electric charges do, sometimes attract and sometimes repel others. Such explanations as have from time to time been offered are far-fetched and by no means convincing. The uni- directional nature of gravitational fields is just one of those stubborn facts that have persistently defied explanation.

22.4: *The Relative Intensities of Electrostatic and Gravitational Fields*
Nearly all inert masses consist of protons, electrons and neutrons. All protons have identical masses and identical positive charges. All electrons also have identical masses and identical negative charges. All neutrons also have identical masses and zero charge. Thus there are con- stant ratios between

146

charge and inert mass. As these ratios are not directly concerned with gravitation they need not concern us here.

But it is generally assumed, rightly or wrongly, that each proton and each electron has attracting as well as inert mass. On this assumption a proton, for instance, is the centre of two fields of which one is electrostatic and the other gravitational. If it is so, one has to conclude that there is a law according to which a motionless electrostatic field cannot occur unless it is accompanied by a gravitational one. (The word 'motionless' is necessary because the field that moves with the velocity of light and is coupled with a magnetic one is not coupled with an attracting one.) One has further to conclude that the ratio between the intensities of the electro- static and gravitational fields around an elementary particle must be one of a few constant values.

As already mentioned electrostatic fields, like gravitational ones, may be regions in which the geometry of space-time is non-Euclidean. It is indeed difficult to imagine what else they could be. But if they are, their geometry must be of a fundamentally different kind from that of the gravitational field; for it does not affect the movement of inert masses; the path of a neutron is the same whether it passes through an electrostatic field or not. Correspondingly the path of a proton or an electron is not influenced by a gravitational field except by virtue of its inert mass. And yet, according to the generally accepted view, gravitational and electrostatic fields invariably occur in association around every charged particle. By what principle, one is led to ask, should two quite different (and non-interacting) kinds of field be coupled in this way? By what principle is the ratio of the intensities of the two fields limited to certain constant values? Existing knowledge offers no hint of an answer to these questions.

These are the questions that are being asked by those who are seeking a unified field theory. They will not be further discussed in these pages. But there is a further question about the ratio of the charge and attracting mass that is, by traditional theory, attributed to elementary particles and this question does have a place in the present inquiry.

Why, it may also be asked, are the two field intensities of such widely different orders of magnitude? The electrostatic field around a proton is large enough to be readily measurable. But its attracting field, if it has one, is so weak that vast numbers of elementary particles, the numbers that make up a sizeable leaden sphere, are needed to form a field of measurable strength.

It has been suggested that gravitation may be a differential effect; and if a gravitational force were indeed simply the difference between a push and a pull, the pull preponderating very slightly, one would, of course, expect the force to be very weak. But that suggestion is pure speculation and is not supported by any evidence. That the two fields associated with the same particle should have such enormously different intensities has yet to be accounted for.

22.5: *Gravitation is Uncontrollable*
A further puzzling feature of gravitation is that it is, in an absolute sense, uncontrollable. This cannot be said about other natural phenomena. Even radio-activity can be influenced by nuclear bombardment.

One can direct a beam of light in any desired direction. One can vary its intensity and colour, reflect and refract it, break it down into the component wave-lengths of its spectrum, polarize it. One can so control temperature that a given substance may be nearly as cold as the absolute zero or hotter than the sun. One can insulate a charged conductor and thus prevent the charge from having an influence on bodies in its vicinity. One can shield a magnetic pole and thus exclude magnetism from places where it would interfere with an experiment. One can conduct electricity along any desired path, vary at will its magnitude, vary its rate of change, switch it on and off. One can arrange and rearrange the chemical elements into a virtually unlimited number of different compounds, accelerate or retard the speed with which they react, influence the physical properties of substances. One can, in short, obtain in any given place any desired amount and quality of light, heat, sound, electricity, magnetism, chemical property.

None of these things can be done with gravitation. One cannot direct, reflect or refract it. One cannot vary its intensity. One cannot conduct it. One cannot insulate or shield it. One cannot switch it on or off. Why not?

We have become so used to the uncontrollability of gravitation that we tend to take it for granted and do not often ask questions about it. When a writer of science fiction introduces into his story the notion of a substance that intercepts gravitation or of a machine that controls it, we may smile at the ignorance of those gullible readers who accept such an impossible notion uncritically. But we should rather reflect on our own ignorance; for there is not a scientist who can say why such an insulating substance and such a controlling machine are but figments of an exuberant imagination, never to be implemented in the world of reality.

It is true enough that one cannot control the weather either. But its uncontrollability is of a different kind from that of gravitation. The reason is that the circumstances by which it is influenced are on a scale beyond man's mastery. Changes in the weather are brought about by the movement of vast quantities of air, changes in the ionization of hundreds of cubic miles of the upper atmosphere, changes in atmospheric pressure, sunspots, in fact by all kinds of occurrence that, though uncontrollable in practice, are not so in theory. If we disposed of unlimited resources, we could control the weather.

Even though meteorology is not a subject on which much research can be done in the laboratories of our scientific institutions, it is a suitable subject for study in what is sometimes rather fantastically called nature's laboratory.

Many things happen in this laboratory to light, heat, sound, electricity, magnetism, chemical compounds, the nuclei of elements. Circumstances, great or small, may cause the same thing to become brighter or darker, hotter or colder, to move faster or slower. The sun, like other objects, undergoes many changes. A part of its surface sometimes experiences a magnetic storm. Its surface is a region of turmoil.

Gravitation, on the other hand, is no more influenced by cosmic than by man-made events. No circumstances ever cause the same body sometimes to exert an observably greater and sometimes a smaller gravitational pull. The inert mass at the centre of a gravitational field may be hot or cold, diffuse or concentrated, at rest or violently agitated, electrically charged or neutral, it matters not. The steady field persists. There are magnetic storms on the sun but the sun is always quite free from gravitational storms. The most shattering cosmic catastrophe does not send as much as a faint tremor through any gravitational field. Gravitation is not observed intermittently, but continuously, so that we never for a moment lack evidence that there is such a phenomenon. We feel its steady influence when we walk and when we lie down to sleep, when we raise an arm and when we swallow our food. Gravitation is literally the only natural phenomenon to assail our senses insistently and without interruption.

True, it is just possible to argue that the above statement may not be quite strictly correct. If it could be proved that the attracting mass, M_g, of a body increases with its velocity according to the law by which the inert mass, M does so, one could control gravitation, be it to a limited extent, by changing the velocity of the source of the field. If the leaden sphere used to measure G were spun, a different value of this constant would be obtained than if the sphere were at rest. If the sphere were heated, moreover, the component molecules would have a greater velocity. It is known that they would also have a greater inert mass. If they had a greater *attracting* mass as well, a neighbouring sphere would be attracted with a greater force.

But I have shown in Chapter 21 why such a conclusion is not implicit in relativity theory. Those inclined to disagree with me about this should remember that the relativity equations seem peculiarly difficult to interpret in physical terms, which explains why experts sometimes differ about their correct interpretation. In such interpretations as I have found there has been no attempt to discuss the relation between attracting mass and the other two kinds. The problem of understanding this relation has not been solved by relativists; it has not even been tackled, it has ignored.

22.6: *The Gravitational Field of a Body is Changeless*.....................................

To say that gravitation is uncontrollable, even in nature's laboratory, is to say that the gravitational field of a given body is changeless. In this it is different from most other natural phenomena. These do change with circumstances. Even the electrical field of a proton or an electron does so. For it may be converted into a photon .This difference between the electro- static and the gravitational field may, incidentally, explain why all attempts at finding a unified field theory have failed.

But there is one other class of phenomena that are as changeless as gravitation. The emission from a radio-active isotope is also a constant quantity, no matter what is done to the emitting body. A given lump of radium produces a known, constant amount of radiation and has a known, constant gravitational field. But every other property that one may assign to it varies with circumstances.

However, there is one significant difference between the changeless radio-activity per unit mass and the changeless gravitational force per unit mass. The former is attributed to a change in the material, namely to the disintegration of atoms; but according to the traditional view the latter is not attributed to any change at all.

The changelessness of the gravitational field is, incidentally, the reason why there was no god of gravitation in the mythology of the ancient Greeks. They believed that all natural phenomena that they could recognize as such were controlled by some mythical being. All events depended, in their mythology, on someone's conscious will just as events in our laboratories do. The notion of nature's laboratory would not have been remote from the Greek attitude. The Greeks saw the sea sometimes tranquil and sometimes in violent turmoil and attributed the changes that it underwent to the work of the sea-god Poseidon. They were alarmed by a raging storm with flashes of lightning and peals of terrifying thunder and explained it as a consequence of the anger of the god Zeus. They saw the fruits of the field grow and ripen ready for the harvest and said that the process was guided by the goddess Demeter. When a tree that began as a sapling grew into a thing of grace and splendour, they concluded that a charming dryad was the cause of the miraculous transformation.

But it was only changes, not steady states, that they attributed to their gods. If the weight of things were to change from time to time, so that, according to circumstances, a stone or a spear was sometimes heavier and sometimes lighter, the ancient Greeks would, no doubt, have included a god of gravitation in their pantheon. But the weight of a thing is the only property that never changes. So there was no such god. Had there been he would have had an idle time.

The remark is not as far off the point as one might think, for the reason why there was no god of gravitation in the Greek pantheon is fundamentally the same as the reason why there are no chairs of gravitation in our universities. A phenomenon on which one cannot conduct experiments, on which experiments are not even conducted in nature's laboratory, does not lend itself to much scientific study. Like a god of gravitation, a professor of that subject would have an idle time, and a dull one.

22.7: *Gravitational Effects do not seem to be Associated with Expenditure of Energy*
It has already been pointed out that the sun expends vast quantities of energy in sending its rays into space, but that, according to the traditional view of gravitation, it does not expend any energy in keeping the planets to their orbits. This important difference between gravitation and other phenomena deserves some discussion.

One can speak of the energy in an electrostatic field. This is produced when electrical charges are separated and the energy consists in the work done to separate the charges. When the charges are allowed to fall together again, the energy is released and appears in some other form.

The gravitational field is not analogous. While energy is expended in creating the electrostatic field, none is expended, according to the traditional view, in creating the gravitational one. It is claimed instead that this field just is. When, again, an object falls under the influence of the electrostatic field, when, for instance, a charged particle falls down the potential gradient, the field is weakened. This is the reason why one says that the energy is stored in the field. But it seems to be meaningless to say that energy is stored in the gravitational field. One can speak of the energy stored in the water of a mountain lake, for this, when released, flows over stones and pebbles and the friction converts this energy into heat. But the field that causes the water to flow down the valley is not in the least reduced in the process; the earth's hold on the moon is as strong after, as before, the water has flown out of the lake.

22.8: *Action at a Distance*
The earth is in one place and the meteor that is accelerated by it is in another. Yet we attribute the acceleration to the earth. So the earth exercises an influence in a place where it is not. How? This is yet another of the facts about gravitation that have to be explained.

The problem is known as that of action at a distance. That it presents itself in Newtonian mechanics is obvious, but strangely enough it has sometimes been denied that it also presents itself in relativity theory. However, the only difference is in the thing on which the earth is assumed to act.

According to Newtonian mechanics the earth acts directly on the stone, though it be from afar. But according to relativity theory the action of the earth is indirect. The earth is claimed to act, not on the stone, but on space; it causes this to be curved. It is the space that, in turn, acts on the stone. So action at a distance is presumed in both theories; in the one it is between the earth and the stone, and in the other between the earth and the part of space where the stone is.

The expression 'action at a distance' is ambiguous. It could mean a kind of action that does not puzzle us at all. When a person sends a letter to someone in another town and asks him to do something he can be said to be exercising an influence in a place where he is not. When a shell is fired out of a gun and explodes some miles away one can say that the gun is exercising an influence at a place where it is not. One can say the same of a lamp in a photographer's studio. The lamp is in one place and the photographic plate on which the light falls in another, and yet the lamp exercises an influence on the plate.

In these examples the action at a distance is effected by means of a transfer of energy; something moves from the one place to the other. Action is between the writer and the letter that he is writing in one town and between the letter and the recipient in the other. It is similarly between the gun and the shell in one place and between the shell and the building destroyed by it in the other place. It is between the lamp and the light rays in one part of the photographer's studio and between the light rays and the photographic plate in another part. In all these cases nothing separates the cause and its immediate effect and the distance to the final effects is spanned by something that moves from the one place to the other, be it a letter, a shell, or light waves. Such movement involves the transfer of energy from the one place to the other and so one can speak of the kind of action at a distance that is effected by a transfer of energy.

This is not usually understood by the expression 'action at a distance'. What is meant is action without an accompanying transfer of energy. It is so understood in the generally accepted view of gravitation, both in Newtonian mechanics and in relativity theory. At the moment when a stone falls it is not assumed that something leaves the earth and travels towards the stone. It is assumed in Newtonian mechanics that the falling occurs without the simultaneous transfer of anything from the earth to the place where the stone is.

The same assumption is usually made in relativity theory. It is assumed that the mere presence of the earth causes space in remote places to be curved. It is not assumed that to maintain this curvature anything leaves the earth and produces its effect at the remote place. The earth is not said to be depleted of anything by creating this curvature around itself. Once again the analogy to radiation and other phenomena fails. When there is action at a distance by the transmission of energy the action is the result of something that happens. This is easy to accept. But it is not so easy to accept the notion of action at a distance as the result of something that merely is. A radiating body loses energy when it acts at a distance; but the traditional view is that the earth acts gravitationally at a distance without losing anything, be it repeated.

Here the mathematician is in an enviable position. He can write down an expression for action at a distance and is not called upon to give it a physical interpretation. But the physicist is in a quandary. His task is to give meanings to the symbols with which the mathematician has provided him. If the symbols say that there is action between distant places without transfer of energy from one place to the other he is puzzled.

22.9: *Summary*

The questions discussed above remain to remind us that we do not yet know as much about gravitation as we should like to know. For the sake of convenience let them be repeated here. They are:

(1) Can gravitation be brought into association with the main body of physical science?

(2) Why do inert and attracting mass always occur in association? The question can take the alternative form: Why does an accumulation of inert mass cause the surrounding space to be curved?

(3) Why does gravitation seem to contradict the Principle of Stabilization according to which every process in a self-contained system tends to reduce its cause?

(4) Why do gravitational fields always have the same sign?

(5) Why is there a limited number of constant ratios between unit electric charge and the inert mass of the elementary particles? Why, moreover, are the gravitational fields attributed to such particles vastly weaker than the electrostatic ones?

(6) Why is gravitation uncontrollable?

(7) Why is gravitation the only phenomenon that does not seem to require a transfer of energy?

(8) How can one reconcile what is known about gravitation with action at a distance?

These are, I venture to claim, the kinds of question about natural phenomena that scientists find it profitable to ask. Answers to this kind of question usually lead to new insight into the nature of the physical world. At least some of these questions have not been ignored in the past, perhaps. But they have certainly not been asked with the insistence that they require. It may be that the prospect of finding answers has seemed to be too remote to justify the effort. But I hope to make it clear in the following chapters that such defeatism is not warranted. There is an excellent prospect that satisfactory answers to very nearly all the above questions can be found by the same procedure that has served for answers to questions about the spiral nebulae. No further hypothesis will be needed than that of Minimum Assumption.

Chapter 23

Misplaced Efforts at Understanding

23.1: *Symmetrical Impermanence is Difficult to Accept but not to Understand*

The many unsolved problems about gravitation reveal, it has just been pointed out, a serious gap in our knowledge of the physical world. It will be shown here that there is a good prospect that a bridge can be provided to span the gap provided the traditional hypothesis about gravitation is replaced by another one. This will be shown to meet Occam's razor better than the traditional one does and thus to be justified both by the criterion of minimum assumption and by that of maximum explanatory power. But its foundations will be different from what one might perhaps expect.

At one end of the span of the bridge the foundation will be the Hypothesis of the Symmetrical Impermanence of matter and at the other end the relativistic notion that space has physical properties. In other words it will be shown that a gravitational field around an accumulation of inert mass can be inferred as the effect on the geometry of space of the continuous origin and extinction of matter.

If both these basic notions were fully established and understood, I could proceed forthwith to build the bridge of logical reasoning that spans them. But it is not so, and anyone who rejects either of the two notions will not be inclined even to consider what inferences can be built on them. Similarly, anyone who does not properly understand either notion is likely to reach a wrong conclusion. Hence I have no choice but to allow a little time for examination of the difficulties inherent in both notions before I can begin to show how a gravitational field can be inferred from them.

Understanding a notion and accepting it are by no means the same thing. I believe that many people can understand the notion of the con- tinuous origin and extinction of matter quite easily, but I have found also that many cannot accept it. On the other hand, a majority of physicists accept today the notion that space has physical properties, but even physicists find the mathematical basis of this conclusion difficult to understand. This distinction has to be faced and appreciated. It is with the big subject of *understanding* that this chapter is chiefly concerned. But it will be helpful if I first give passing attention to the fact that the notion of Symmetrical Impermanence is difficult to accept. If we cannot remove the difficulties of acceptance as well as of understanding, we must at least try to come to terms with them.

What makes Symmetrical Impermanence difficult to accept is that the concept is new, unfamiliar, at variance with tradition. To some it may also appear unattractive, for if matter itself is impermanent the very foundations of science and the universe may seem to be threatened. The resultant sense of insecurity may cause some to feel uneasy. For a physicist to be influenced by such emotional considerations is, admittedly, reprehensible. But we are all human and cannot quite avoid judging a notion at least partly by the criterion of attractiveness, even when we know that this criterion has no place in science.

If, however, objections to Symmetrical Impermanence can be urged in plenty, I do not think that incomprehensibility is among them. Incredible' is a more probable epithet from those who do not like the notion. It may be difficult to believe that an elementary component of the material universe originates and becomes extinct without cause, but the idea is not difficult to understand.

For Symmetrical Impermanence it is thus acceptability and not comprehensibility that needs to be achieved before attention can be given to my new theory of gravitation. How can this be done? Acceptance cannot be urged by reference to authority, for orthodox physics accepts, as though it were proved and irrefutable fact, that the future duration of every component of the material universe is infinite and that its past duration dates from some moment that occurred between the beginning and the completion of the Creation. To suggest, as I have done, that there is a half-life for matter in the sense in which there is a half-life for a radio-active element is today regarded by authority as a heresy.

The justification for my suggestion is, as has been explained in earlier chapters, not to be found in tradition. It is that it meets the two criteria of minimum assumption and maximum explanatory power, as has been demonstrated already. It has been shown by closely reasoned steps that one can infer many observed facts about the nebulae from Symmetrical Impermanence and not from any of the more orthodox, rival hypotheses about the duration of matter. This alone should suffice to gain serious consideration for Symmetrical Impermanence; and a further justification will, I am claiming, be found in its ability to explain those most recalcitrant phenomena connected with gravitation that were enumerated in Chapter 22.

23.2: *Relativity is Difficult to Understand*
But if Symmetrical Impermanence is unorthodox relativity theory is, as I have said above, a part of established physical science. Early disbelief has had to yield to acceptance. Incredible' is no longer the right epithet for the statement that space has physical properties. We have to believe the experts, whether we

find it easy to do so or not, when they tell us that space-time can be curved, that it forms a single, indivisible physical concept, that space is not a container but a constituent of the physical universe, that matter and space are conceptually inseparable, that space expands. But I fear that 'incomprehensible' still remains a justifiable epithet for all this. The meaning of what the relativist tells us about the nature of space is as hard to understand today as it was half a century ago when the theory first startled the scientific world.

This is the situation with which we have to come to terms. To do so something less will suffice than the degree of understanding that a specialist in relativity theory requires. The specialist's services will be required, of course; but not until later. It will be for him to help test the reasoning by which my new theory of gravitation has been reached. But the reasoning must be presented by myself and followed by many who are not specialists before it can usefully be tested, and for this purpose it will not prove necessary to make the same effort at understanding relativity theory by which physicists satisfied themselves half a century ago about the soundness of Einstein's conclusions. (This is fortunate, for those who are today fully conversant with relativity theory are no longer numerous.) But the non-specialist who would follow my reasoning must, at least, appreciate what kind of understanding can be achieved and what kind of effort must be made in order to achieve it.

This would be easy if the word 'understand' always meant the same thing and always called for the same kind of effort. But it is far from so. There are several different kinds of understanding and quite a different kind of effort is required for each. 'Understand' is one of the vaguest of words, and one can only come to terms with the difficulty inherent in relativity theory if one appreciates this. I therefore make no apology for entering once again the realm of the philosophy of science and for delaying for a short while the promised new theory of gravitation.

23.3: *Different Meanings of 'Understand'*
Of the several meanings that the word 'understand' can have, only two have a place in physics. These will be discussed below, but it will be helpful if I first say a little about those meanings that do not have a place in physics.

Psychological understanding is one of these. It has an obvious and important place in the humanistic disciplines. It calls for the kind of effort that one makes when one pictures oneself in the situation of someone else. By this effort one is able to imagine oneself having a similar experience to the other person and thus understanding how he feels and why he acts as he does. It is thus, for instance, that one may achieve understanding of Hamlet's inability to perform the action to which his conscience prompts him.

No one believes for a moment nowadays that understanding in physics is of this kind. But it has not always been quite so. There was a time when it seemed that understanding of the mercury barometer depended on an effort to appreciate Nature's horror of a vacuum. In those days, and perhaps even later, those whose training had been wholly humanistic must have found difficulty in abandoning the notion that rivers follow an urge when they flow seawards.

Teleological understanding is another kind that has no place in physics, although it has an important place in other disciplines. In seeking to understand a thing one often seeks to discover its purpose. To say that one understands why a particular law has been enacted usually means that one knows its purpose. It is similar with many other human actions. To reveal their purpose is to make them comprehensible.

This kind of understanding is of decisive importance in engineering. In a well-designed machine every feature serves a purpose and an understanding of the machine follows from knowing what the purpose of each feature is. As, moreover, a living organism can aptly be described as a machine, one should expect the same kind of teleological understanding to have a place in biology. But those who study the mechanisms that operate in plants and animals, and who should be in the best position to appreciate the purpose of the component parts, often deny with some vehemence that the kind of teleological effort at understanding that an engineer brings to his machines applies to a living machine. Which is odd.

In ethics, again, understanding means something not quite satisfactorily included in any of the above categories. The word may be applied in ethics to the kind of understanding that leads to knowledge of the difference between good and evil. In the days when some things and substances were thought of as good and others as bad this kind of understanding seemed to help towards an appreciation of the physical world. It seemed easy in those days to understand the smoothness and beauty of planetary orbits when one thought of the planets as among the very best of the good things.

Further, there is aesthetic understanding. It is very different from the other kinds and calls for a different mental effort. We make this effort when we try to understand a piece of music that is written in an unfamiliar idiom. It is only after listening repeatedly and with concentrated attention that we succeed in recognizing the pattern of the notes, the form of the melodies, the structure of the whole, the relation between the harmonies, the nature of the progressions, the pulse of the rhythms. Until this has happened the piece appears as a meaningless jumble of unrelated noises. But when we have appreciated the order we say that we understand the piece.

It would be idle further to elaborate the catalogue of meanings of the word 'understand' that have no place in physics. It is arguable that some of the above uses are really misuses and are to be regretted. But that would only confirm the contention that only too often insufficient care is taken to define what is meant by the word.

I should not have thought it necessary to insist on this if the word always meant the same thing in physics. In that case the warning that care must be taken would be redundant for there would be no risk that an effort at understanding would ever be misplaced. But in fact the word may have one of two quite different meanings, both appropriate to physics, and it does at times happen to all of us that we inadvertently direct our effort towards the one when it ought to be directed only towards the other.

As names are needed I shall call these two kinds respectively 'deductive' and 'representational' understanding. Each will now be described and discussed in turn.

23.4. *Deductive Understanding*
This calls for an effort at deductive reasoning and the method is applied to statements in mathematics as well as to those in physics. The word is used in this sense when one says that one understands the rule that relates the squares on the sides of a right-angled triangle to each other. To say this means that one can follow the reasoning by which the statement is arrived at. But this kind of understanding is not confined to mathematics.

To say that one understands the principle of the lever, for instance, does not mean that one understands the psychology of this device, or the purpose that it serves, or its moral status, or its aesthetic significance. It simply means that one knows how to infer, by deductive reasoning, the ratio of the forces that are simultaneously applied to its two arms. Similarly, a claim to understand the parabolic path of a projectile simply means that one can show by deductive reasoning how the forces that act on the projectile cause successive positions to lie on a parabola.

The same kind of deductive reasoning was brought by Newton to the elliptical paths of planets. It is brought to the rise and fall of the tides, to the movement of mercury in a barometer, to the helium synthesis from which the sun derives the energy that it radiates, to the formation of a cloud from super-cooled water vapour. I have tried to show in preceding chapters how one can understand the occurrence and structure of the nebulae by making the same kind of deductive effort.

When a relativist assures us that he understands why space has physical properties, he means that he has achieved this kind of deductive under-standing. He has successfully made the necessary effort and has been able to infer from certain known facts those properties of space that cause the track of an inert mass in space-time to be curved. The reasoning by which he arrives at the conclusion is, of course, far more difficult than the reasoning that leads to an understanding of the principle of the lever, but it is a quantitative and not a qualitative difference.

All this may be called rather obvious. I think it is. But it leads to an important question. When those of us who are not mathematical experts say that we cannot understand what the relativists tell us about the physical properties of space, do we mean no more than that we are unable to follow the reasoning by which these properties have been inferred? Should we be satisfied that we understood relativity if we knew how to set up and solve the equations? We know that mathematical ability is the key by which the door to the relativist's world is opened. It is the door labelled 'deductive understanding'. But is necessarily this the door that we wish to pass through? Would deduction give us the kind of understanding for which we crave? Do we not suspect that if this door were opened for us it would only lead on to a narrow passage with another closed door at the far end?

The metaphor must not be pressed too far. But this is, I think, how many of us feel about relativity and much else in modern physics. We are prepared to take the results of deductive reasoning, mathematical or otherwise, on trust. If we could be gently guided through that reasoning, it would make no great difference to our state of mind. Our uneasiness about those modern discoveries in physics that are so difficult for us to understand does not arise from any doubt as to their logical justification. It arises from our inability to achieve a different kind of understanding. This is the kind that I propose to call representational. It is the key to the door bearing this name that we are often seeking; and we should go on seeking it even though the door labelled 'deductive understanding' stood wide open before us. But I shall show in a moment that we must often seek in vain.

23.5: *Representational Understanding*
Representational understanding has nothing to do with deduction but calls for an effort of the imagination. It contributes substantially to the work of both the physicist and the engineer and is often complementary to deductive understanding.

Both kinds are used, for instance, when we attempt to understand the principle of the lever. We do not then rest content with the reasoning by which the ratio of the forces on the two arms is deduced. With an effort of the imagination we also conjure up this device before the mind's eye. We should not be satisfied that we fully understood the principle unless we could add a mental picture to the algebraic symbols.

In the example of the lever the representational effort is slight. On other occasions it may be considerable. One can achieve deductive under- standing of gyroscopic action with the help of mathematical symbols. Some of these are operators defining a direction; and in three-dimensional space the direction for any one of them is at right-angles to the direction of the other two. But the logic of the symbols, irrefutable though it be, leaves us unsatisfied. If we are to feel happy about our understanding of the gyroscope we must also succeed in imagining the spatial relations of the system; we have to represent to the mind's eye the forces and movements that occur in all three dimensions. It is not altogether easy.

The slide valve system of a steam engine provides an even more cogent example. The components of the system include devices known as eccentrics, sliding members that move in a complicated way relative to each other, ports that are open to the passage of steam at certain moments and closed at other moments. These ports begin to admit steam to the cylinder when the piston is in a certain position and cut the steam off when it is in another position. In yet another position of the piston they allow steam to be discharged from the cylinder. Some of the various moving parts are in circular, others in reciprocating motion. The complexities of the system extend in all three dimensions.

For this intricate piece of kinematics the deductive effort at understanding is rather easy, the representational effort is quite difficult. But no engineer would claim to understand the action of a slide valve unless his representational effort had proved successful.

23.6: *Representational Understanding is Sometimes Impossible*
It is only natural to wish to understand every statement in physics in the representational as well as in the deductive sense. As has been seen from the examples of the lever and the gyroscope, we often succeed. But it is not always so. The human imagination has its limit as regards representational understanding and many accepted statements in physics lie beyond this limit.

Statements about the electron provide an example. When this particle is mentioned we try to visualize it, and each of us forms, no doubt, his own different picture. Some may think of it as a hard little sphere, others as more like a ball of loosely-tangled wool; to some its surface is perfectly smooth and featureless, to others rough enough to make the spin perceptible; to some the boundary is definite, to others the electron carries with it a sort of misty nimbus; to some it is black, to others yellow.

On occasion a fact forces itself on our attention that causes us to revise whatever private picture we may have formed. The electron consists of waves, we are reminded, and its field extends into its surroundings: forthwith the little hard sphere turns in our imagination into the tangle of wool and is surrounded by the extensive misty nimbus. But the electron also finds room in the tiny atomic nucleus: the hard sphere comes back and the nimbus is shed. The electron spins, we recall: and we imagine some surface markings to feature successive angular displacements. But one electron is so like every other that it is meaningless even to say that two electrons ever change places: gone are the surface markings.

So it goes on. No one picture is any nearer the truth than any others. Any picture that the imagination can conjure up must be false. And yet, how hard it is to prevent the imagination from intruding. What would the electron be like, we feel impelled to ask, if it were magnified to the size of a tennis ball? It would then *have* to be either hard or soft, rough or smooth, provided with either a clear or a fuzzy outline, be either black, or white, or coloured. Which of all the conceivable possibilities would it be?

If we are honest, we have to admit how difficult it is always to remember that these questions are silly. If we have adequate understanding of the deductive kind we know they are silly; we know that the association between the size and the other properties of the electron is not the same as it is for tennis balls. While a tennis ball of non-standard size may have all the same properties as a tennis ball of standard size, an electron of a different size would not retain a single one of its properties.

Knowing all this most of us still persist, be it confessed, in forming our mental picture of the electron. Our urge towards understanding in the representational sense is so strong that we like to pretend to ourselves that we have achieved it even when we know that we have done nothing of the sort. We then have no choice but to imagine a picture that we know to be false.

Even such poor comfort as may be obtained by this self-deception is denied us when we are seeking to understand what relativists have discovered about the physical properties of space. Space-time, expanding space, curved space, space with any features at all defy the imagination. We cannot conjure up even a false picture of it!

23.7: *Objectivity is the Enemy of Representational Understanding*
The reason for this inability is as simple as it is profound. When one makes an effort at representational understanding one tries, it has been pointed out already, to represent something to the imagination. One tries to see it with the mind's eye, to hear it with the mind's ear, to feel it with the mind's sense of touch. One associates it, in other words, with one's own personal, subjective capacity for experience. But it will be shown in a moment that the physicist's aim is to find aspects of reality that cannot be thus associated. What he can say about these aspects is independent of any particular observer; it is truly objective. This is why it is axiomatic that objectivity, in the physicist's sense of the word, and representational understanding are irreconcilable.

This very fundamental fact about physics needs to be elaborated a little.

In physics colour can be defined in terms of wave-length, as can also pitch of sound; temperature can be defined in terms of the average momentum of the molecules in a substance and taste in terms of chemical constitution. Why is this done?

The answer can take a variety of forms and one of them is to say that this way of defining properties gives them a more universal meaning. To say that a surface is red has meaning only for a person whose eyes can distinguish colours; it means nothing to a colour-blind person. But the information that the surface reflects radiation of such and such a wave-length conveys something to the colour-blind man, and indeed as much as it does to the one who has had the experience of distinguishing between a red and a green surface.

The same universalizing process extends into all aspects of the objective world. The physicist can tell us things about sound that have a meaning for a deaf man, some things about light that have a meaning for a blind one. He can say things about temperatures so high that no man could experience them and live, about particles too small to be perceived by any of our sense organs. The physicist thus aims at finding statements that would take the same form if people had different sense organs and a different nervous system.

Therein lies the justification, or at least one of the justifications, for the generalizations of physics. They aim at abstracting from the whole world of reality that part that is truly objective in the sense that it is independent of any particular observer. What can be said about this aspect must, by definition, have the same meaning for every receiver of the information whatever his physiological and psychological constitution may be. For this reason the statements that are regarded in physics as the greatest generalizations are those said, in technical language, 'to eliminate the observer'.

Such statements are, of course, incomplete, for the observer is still there; subjective experience is a reality that cannot be annulled by saying nothing about it. But the incompleteness is deliberate. This conceptual distinction between objective and subjective reality is one of the greatest achievements of our Western civilization.

Having made it we have to appreciate the consequences. One of them is that the widest generalizations in physics cannot be expressed in terms of sense data. If a statement about light means as much to a blind man as to one who can see it cannot refer to the experience of seeing and it cannot be presented to the visual imagination. If the observer has been eliminated, one cannot regard him as still using his organs of sense perception. The objective generalizations that the physicist seeks to find are just those that the observer could not become aware of by using his five senses. To represent something to the mind's eye is to imagine that one is observing it, that one has not been

163

eliminated. Hence any effort at representational understanding of truly objective statements in physics must fail by the nature of the subject.

All this means that in some branches of physics there is no key to representational understanding. We are separated from it, not by a door, but by a blank wall. With his acute deductive understanding the physicist has discovered things that are, and must always remain, beyond this kind of understanding. It is an understatement to say that efforts at representational understanding are then misplaced. They are also misleading. They are efforts at replacing true concepts by false ones.

Chapter 24

Space

Space 24.1: *Inferential and Observational Evidence*
What has been said in the last chapter about the impossibility of understanding some of the notions with which physics is concerned in the representational sense of the word 'understand' applies forcibly to the notion of expanding space. It defies every effort of the visual imagination. But we have, nevertheless, two kinds of evidence for it: inferential and observational.

In physical science we are rarely satisfied about the validity of any statement unless it has the support of both these kinds of evidence. The inferential evidence for the performance of a machine is, for instance, provided by calculations made before the machine is even in the blueprint stage. We regard it as nearly, but not quite, conclusive. The observational evidence is provided by tests made with the machine after it has been built. If they confirm the calculations, we are satisfied. Similarly, both kinds of evidence are usually provided by a teacher who is lecturing on one of the generalizations of physics. He first deduces the generalization from first principles by making drawings and calculations on the blackboard and then confirms what he has just inferred by an experimental demonstration.

If our reasoning were infallible, one of the two kinds of evidence alone would suffice. But we cannot be thus sure of ourselves. When there is only inferential evidence, one is justified in asking whether this may not be based on false premises or faulty deduction. When there is only observational evidence, this may be open to different interpretations; we are justified in doubting whether one may draw the stated general conclusion from a particular case. But when there are both kinds of evidence, each supports the other.

It is only because the notion of expanding space has this dual support that it is widely, though not perhaps universally, accepted. It will suffice here to refer quite briefly to the two kinds of evidence for it.

The inferential kind was provided first, and by those who had mastered the relativity equations. They explained that a cosmological model of which the volume did not change with time would be unstable. In the sense in which they used the word 'unstable', a model that resembled actuality would have to change its volume. This did not prove that space expands. The inference would have been equally compatible with a model that contracted, which it

165

could, of course, not have been doing for an indefinitely long time without having disappeared. Hence the conclusion that space was, in fact, expanding was first arrived at by reasoning alone.

The observational evidence for the same conclusion is well-known. It is provided by the red shift in the spectrum of the light from distant nebulae. This shift is interpreted as a Doppler effect and is attributed to a recession of the nebulae from each other. The magnitude of the shift is found to a close approximation to be proportional to the distance of the nebulae. It is the effect that was predicted by the inferential evidence.

When there is such good agreement between prediction and observation, it is usual in physics for the conclusion to be accepted without further question. If it was not so for the notion of expanding space, the reason is not far to seek. This notion cannot be understood representationally any more than the notion of the electron can, and it is only natural to dislike a conclusion that one cannot fully understand. From dislike to rejection is but a short step, so it is not surprising that some rather desperate attempts have been made to find an alternative explanation of the red shift. It has been done, of course, in the name of scientific caution. But the degree of scientific caution with which a new idea is greeted is some indication of its unattractiveness.

Should a convincing alternative explanation of the red shift be found the situation would be that inference supported the notion of expanding space while observation failed to support it. The calculations made by relativists would then have to be re-examined. It really ought to be done simultaneously with the search for an alternative explanation of the red shift. If this has not happened, it is probably because the mental effort or understanding the calculations and examining them critically is considerable. It is far easier to invent an hypothesis by which to explain the red shift or explain it away.

But unless both the inferential and the observational evidence are effectively shaken, the wisest course is to come to terms with the notion of expanding space, whether we find it attractive or not. And it has so far not been effectively shaken. So we must, I am afraid, be content to do with this notion of expanding space what we have had to do with the notion of the electron: form our private mental images of it from time to time; but always remember that these images are wrong. Experience with the electron has shown that we are rarely misled by doing so, provided we recognize the inadequacy of the images and are prepared to replace them by others if and when occasion demands. Experience has also shown that images, false though they be, are sometimes helpful, sometimes even indispensable.

24.2: *How to Interpret the Notion of Expanding Space*
Thus forewarned let us try to reach, if not representational understanding of, at least some valid statements about, expanding space.

One sometimes says that a fugitive from justice puts space between himself and his pursuers. One does not mean the expression to be taken literally. One only means that the fugitive is running faster than those in pursuit. When he does this he does not create new space but only causes a larger amount of existing space to separate him from the pursuers.

If the fugitive could literally put space between himself and his pursuers, he would not need to run away from them. He could sit down and smoke a cigarette while he put enough space in front of those who were trying to catch him to make sure that they never got any nearer. If he did this he would not be moving past objects in existing space. He would not be moving at all.

It is in this literal sense that, in an expanding universe, space originates between us and every distant nebula. While the fugitive from justice is getting further from his pursuers, he is also getting nearer to the house in which he hopes to hide. But while our galaxy is being caused by the expansion of space to get further from all other galaxies, it is not being caused to get nearer to anything.

In this there is a significant difference between changes that result from the operation of forces between bodies and those that result from the expansion of space. So long as there are forces, some things get nearer to others; they overtake other things. But when space-expansion alone determines distances nothing ever gets nearer to anything else; there is no overtaking; there is not even movement.

It is this last conclusion that makes the notion so difficult to understand. If things get further apart they must, we are inclined to reason, move relatively to each other. But we have to appreciate that this is false reasoning. Let me show why as clearly as possible.

Two nebulae, A and B, have been observed and both show the red shift. One of them, A, is in the part of the sky called 'north' and B is in exactly the opposite direction, the part called 'south'. When we are thinking only of A, we may make one of two statements:

(1) The distance between our galaxy and the nebula is increasing.
(2) Our galaxy and the nebula are moving relatively to each other.

We may be inclined to think that these two statements have identical meanings; and so they would have in many contexts. But if we attribute the red shift to the expansion of space we have to conclude that they mean different things and that, while (1) is correct, (2) is wrong.

This emerges when the implications of (2) are examined. To say that our

167

galaxy and the nebula A are moving relatively to each other may mean that both are moving or that one is at rest while the other is moving. But it must mean that at least one of the two bodies is moving.

If this were our own galaxy, it would be moving away from A, *ie* southwards. But when we observe nebula B we have to conclude that, if our galaxy moves at all, it must be *away* from B and northwards. A corresponding conclusion would be reached if we used a nebula in any other part of the sky as an indication of the direction in which our galaxy was moving. Wherever our choice fell, it would always cause us to say that we were moving away from the observed nebula. To say that the expansion of space is causing our galaxy to move is to say that the movement is in all directions at once! The correct interpretation of the red shift is, in other words, that our galaxy is not moving at all relative to any other nebula.

Are we then to take the view that we alone are at rest and that nebulae A and B, together with all others, are moving relatively to us? Are we to adopt the old egocentric universe in which we are located at a centre from which all effects radiate?

This, we know, cannot be. An observer on any other nebula would have the same experience as we ourselves here. It would be just as impossible for him to state the direction in which his nebula was moving. He could not say that it was moving relative to space in such a way that space, at one moment in front of it, was behind it at the next moment. He would say that his nebula was not overtaking anything, not even empty space; that it was not moving in any direction; that it was at rest.

We are thus obliged, whether we like it or not, to accept the odd notion that in expanding space the distance increases between objects that are all at rest relative to their surroundings. If the nebulae all seem to drift away from each other, this cannot be attributed to anything that happens to the nebulae but only to what happens to the space between them.

It is amusing and instructive to try to find a mathematical formula that would express a force of repulsion between nebulae in an expanding space; for the attempt is bound to fail. The velocity of recession as viewed in one particular direction is:

$dl/dt = Hl$

where l is distance and H is Hubble's constant. It has a value of 185 kilometres per sec per megaparsec. The acceleration is:

$A \, d_2l \, / \, dt^2 = Hdl \, / \, dt = H^2 \, l \, ... (24a)$

Here A is the acceleration of one body relative to a single selected other one, but not the acceleration of any body relative to all other ones, which would always be zero.

Force is the product of mass and acceleration. If our galaxy were receding under the influence of a repelling force, we should therefore give this the value:

$$F = mA = mH^2l \dots\dots (24b)$$

where m was the mass of our receding galaxy. But this would be an absurd conclusion. According to equation (24b), the force exerted by a body on . any other one would not be proportional to the mass of the repelling body ; but to that of the repelled one.

Such a conclusion cannot, of course, be reconciled with the known law of gravitational attraction. Consider two bodies with the respective masses m_1 and m_2. The gravitational force between them, which needs the negative sign as it is one of attraction, is:

$$F = -Gm_1m_2 / l^2$$

Here the force exerted by m_1 on m_2 is the same as that exerted by m_2 on m_1. But if equation (24b) meant anything, which it does not, one would have to express the force exerted by m_1 on m_2 as $m_2 H^2l$ and that exerted by m_2 on m_1 as $m_1 H^2l$. The tiniest repelling mass m_1 would exert a very big force on m_2 if this were big.

Let the distinction between a change of distance that is due to expanding space and one that is due to a force be expressed in a slightly different way. The acceleration of mass m_1 relative to m_2 is $- 1/2G \, m_2 / l^2$, and the acceleration of m_2 relative to m_1 is $-G \, m_1 / l^2$. The total acceleration of the two masses relative to each other in expanding space is:

$$A_{total} = kH^2 l - 1/2G(m_1 + m_2)/l^2 \dots\dots (24c)$$

This expression shows clearly that the relative acceleration occasioned by gravity is dependent on both masses as well as on the distance between them, while the relative acceleration occasioned by the expansion of space is dependent only on the distance between the masses. It is a function of space and of nothing else. For the distance, l, is the only variable in the term that defines the effect of expanding space.

The above equations are, of course, no more than a mathematical way of expressing the conclusion already reached without mathematics that the expansion of space does not cause the bodies in it to move. As they do not move, they are not accelerated by the expansion of space and are not subjected by it to any force.

24.3: *The Origin of Space and the Origin of Matter*
The *expansion* of space could equally well be called the *origin* of space. We may therefore speak of the continuous origin of space as we do of the continuous origin of matter. Are the two origins coupled? Can there be origin of space without origin of matter? Are the two kinds of originating the same process, or are they two separate and distinct processes?

If we could regard space as the container of the physical universe and ponderable matter as the content, there would be no reason why their respective origins should be related. A change in the capacity of the container might well occur without a corresponding change in the quantity of matter contained in it. One could then say that at a certain mass density space was quite full; and that at half this limiting density space was only half-full. There would then be an absolute scale of mass density, as there is an absolute scale of temperature. But no one has ever looked for such a scale; no one has ever had the idea that there could be one; no one believed, even in pre-relativity days, that the expression 'space is as full of matter as it can hold' would have any meaning. Today relativists are able to show why, in fact, it has no meaning.

They are also able to show that the origin of space and the origin of matter are coupled, if not better described as synonymous. McCrea, it will be remembered, has inferred the net rate of origin of matter, namely 500 atoms of hydrogen per cubic kilometre per year, from the observed rate of expansion of space.

Yet here, as often in physics, there is a conflict between the two kinds of understanding. Our misplaced effort at representational understanding insists on retaining the notion of a purely conceptual, featureless space, into which particles of matter could be poured; just as it insists on retaining the notion of a purely conceptual, eventless time along which events are ranged. It is only deductive understanding that tells us that these notions are meaningless and that it is absurd to ask whether space is quite full of matter or only partially full. When visualizing a cosmological model based on the hypothesis (A2) about the Creation, representational understanding tries, misguidedly, to picture the featureless space as existing before the Creation is assumed to have begun, always to be unchanging, and to have been made the recipient of more and more matter as the process of the Creation became more and more complete. Those of us who accept (A3) tend to make the same misplaced effort. But in the battle between the two sides of one's intellect deductive understanding must win and the effort at representational understanding must sometimes be abandoned. It is so here. To speak of the expansion of

170

space is to speak of a space with physical properties and to imply the continuous origin of matter. It is for this reason that the observed red shift provides observational evidence for (A3). (I am only too well aware, let me add, of the conceptual difficulty in the way of postulating the simultaneous origin of space and matter as well as their simultaneous extinction. I am hoping that what is said in Appendix H will help to overcome this difficulty.)

24.4: *The Contraction of Space and the Extinction of Matter*
If the origin of an elementary component of the material universe, of what for short I shall call a particle, is associated with the origin of some space, the extinction of a particle must be associated with the extinction of some space. A region where the rate of origins exceeds the rate of extinctions must then be one in which space is expanding; conversely a region where the rate of extinctions predominates must be one in which space is contracting; and a region in which the rates are equal, i.e. a region at the equilibrium density, must be one in which the extent of space remains constant. This suggests a further means of testing the Hypothesis of the Symmetrical Impermanence of matter by observation.

The average density of the nebulae, including our own galaxy, is much above the equilibrium value. Space within the galaxy must therefore be contracting. Instead of showing a red shift the spectra of the light from stars within the galaxy must therefore show a violet shift. But the shift is proportional to the product of the rate of contraction and the distance of the observed star. This product may not be sufficient to reveal a measurable shift. The Doppler effect occasioned by the relative movement of stars within the galaxy is likely to exceed that occasioned by contraction and to make the interpretation of the readings uncertain.

But there are other ways of testing the hypothesis and one of these can be usefully mentioned here. It would be provided by spectography.

The light from any nebula passes partly through extragalactic space, but partly also through a portion of our own galaxy. According to Symmetrical Impermanence the space within this must be contracting, it has just been said, at a rate given by the excess density within the galaxy over the equilibrium value. The light from a nebula passes, therefore, partly through expanding and partly through contracting space. The latter part of the path will be a small fraction of the whole if the nebula is very distant and a larger fraction if it is close. The red shift that is observed will depend on the arithmetical mean of the contraction over the short path within the galaxy and the expansion over the long path outside.

The nearer the nebula is the greater will be the relative effect of the contracting portion of the path through which the light travels and the smaller the red shift. For near nebulae one should therefore expect, on the average, the value of H to come out rather lower than it would for distant ones.

Observation of the spectra of a large number of nebulae made in the United States by Humason, Mayall and Sandage, as mentioned in Chapter 4, suggests that it may be so. The red shift has been found to lead to a slightly lower value of H for near, than for very distant nebulae. The difference is hardly great enough to be quite conclusive, but it is in the direction that Symmetrical Impermanence would predict.

That contraction of space within our own galaxy reduces the red shift is, however, not the only possible explanation of the observed results. The greater the distance of a nebula the longer its light has taken to reach us. We are therefore observing today the spectrum of light that left the nebula a long while ago. The larger value for the red shift in the light from very distant nebulae might therefore mean that space was expanding more rapidly when this light began its journey than it is now.

The rate might thus vary either with time or with locality. But there seems to be no convincing reason why the variation should be with time, and the continuous extinction of matter provides a reason why it should be with location.

The solar system is by no means at the centre of our galaxy; it is rather far out, though not on the very edge either. So the path of light from a distant nebula always lies partly within the galaxy within a contracting region. But the length of the contracting portion of the path differs with the direction in which the nebula is viewed. It is shortest if it is in a direction at right-angles to our disc-shaped galaxy.

This may provide means of testing the two interpretations. If the variations in the apparent value of H are with time, the red shift will be the same for all equidistant nebulae, irrespective of the direction in which they lie. But if the variations are with locality, the red shift for equidistant nebulae will be less when the nebulae are so situated that their light : traverses a large part of the galaxy and greater if the light traverses a small part. It is conceivable that an analysis of the observations already made would settle the question whether H is constant in space and varies with time, or is, as I am claiming, not a function of time but varies in Space as a function of mass density.

24.5: *Summary*
Let the conclusions reached in this chapter be summarized in a few sentences.

That space expands has been inferred mathematically from first principles and confirmed by observation. Such attempts as have been made to find alternative explanations for the observations have been far-fetched and unconvincing.

The observed recession of distant bodies cannot be interpreted as the consequence of movement of the bodies relative to existing space or attributed to forces of repulsion between the bodies. The only interpretation to fit the facts is that new space is originating continuously between the bodies while these may remain stationary.

The expansion of space is coupled with the origin of matter and no satisfactory hypothesis seems possible according to which the extent of the universe could change while its content remained constant, why, in other words, space could originate and matter not do so.

If origins of elementary components of the material universe must be associated with the expansion of space, the converse must also hold. Extinctions must be associated with the contraction of space. As, according to Symmetrical Impermanence, origins and extinctions proceed side by side, any observed expansion must be a net one and represent the difference between the local gross expansion and the local gross contraction. At the equilibrium density this net difference becomes zero. Where the density is below the equilibrium value there is a net expansion, and it is shown by the expansion of the universe as a whole that the average density for the whole is below the equilibrium value. In any region of high mass density there must be a pronounced net contraction.

Chapter 25

A New Theory of Gravitation

25.1: *To be Valid a Theory of Gravitation must be Based on the Principle of Minimum Assumption*
A considerable portion of the foregoing chapters has been concerned with two hypotheses. The first is that of the continuous origin and extinction of matter, which I have called here the Hypothesis of the Symmetrical Impermanence of Matter. Although I published this hypothesis as long ago as in 1940, the present book is almost the first record of a serious attempt to explore its implications. The second hypothesis is that of the expansion of space, which has already received wide, if not yet universal, support.

It has been shown here that the evidence for both hypotheses is strong and abundant. It is observational as well as inferential. One must not, of course, exclude the possibility that the whole of this evidence may some day be refuted. Alternative explanations may be found for each piece of observational evidence in turn; the Principle of Minimum Assumption, which is the basis of the inferential evidence, may be proved false and have to be abandoned; the arguments that I have been presenting may prove to contain faulty logic; an error may be found in the mathematics by which relativists have inferred the linkage between ponderable matter and space.

Evidence is never so conclusive that anyone should be discouraged from attempting to refute it, and in this instance such an attempt might well lead to a discovery of importance. But the search for means of refuting an hypothesis should include the search for means of testing it, be it by observation or by experiment. Its implications should be worked out and so formulated that one can say: 'If this hypothesis is true, one should expect so-and-so'. One can then make the observation, conduct the experiment, and find out thereby whether the hypothesis has support or not.

This was the method pursued in Part Three for testing the Hypothesis of Symmetrical Impermanence. If this hypothesis is true, it was argued, one should expect to observe spiral nebulae. That they had already been observed made it appear, perhaps, that the hypothesis was justified by its explanatory power rather than by its power of prediction. But there is really no difference. Whether one says that an hypothesis explains or predicts depends, as I have pointed out before, on whether the observation to which the hypothesis refers has preceded or followed it.

Here, in Part Four, three further pieces of evidence will be presented. They are, respectively, gravitation, the occurrence of stars, and their rotation. The first of these will be discussed in this chapter, the second and third in Chapters 27 and 28.

What is to be presented here can be regarded in several different ways. From one point of view it constitutes two new theories, one about the cause and nature of gravitation, the other about the process by which stars are formed. But from another point of view it constitutes confirmation by observation, and in a sense by experiment, of the validity of both Symmetrical Impermanence and general relativity. As both these hypotheses are inferences from the Principle of Minimum Assumption, what is presented here can, more basically, be regarded as a demonstration of the great unifying power of this principle.

25.2: Gravitation as a Consequence of the Extinction of Matter
Let me follow here the line of reasoning by which, in fact, I arrived at the new theory of gravitation. I did not do so in a deliberate attempt to solve the riddles that were presented in Chapter 22. At the time I was but dimly aware of these riddles and had come to regard them as largely beyond the scope of scientific inquiry. What I was concerned with instead was a means of testing the Hypothesis of Symmetrical Impermanence. During my search I found that the behaviour of ponderable matter in the vicinity of a massive body provides such means.

As discussed in Chapter 14, the rate of origins is constant per unit volume and the rate of extinctions constant per unit mass. For reasons given in Chapter 24 and Appendix H the origin of matter and the origin of space occur in association and the extinction of matter and the ex- tinction of space also occur in association.

From these considerations it follows that the gross rate of expansion of space per unit volume is everywhere the same, while the gross rate of contraction is everywhere proportional to the mass density. When the two rates are superimposed, one obtains the net rate of expansion per unit volume, which is positive when the mass density is below the equilibrium value and negative when the mass density is above this value. In other words, a very tenuous region expands and a dense one contracts.

It has already been explained that for this reason the rate of contraction must exceed that of expansion within our galaxy, and must greatly exceed it within every star. The suggestion has been made in Chapter 24 that the contraction might just conceivably be observed for the galaxy as a whole. It would be so if one could measure a reduction in the red shift of the spectrum of the light from distant stars. But the effect would be very slight and might not be measurable. However, a moment's thought will show that one ought to expect the contraction of space within large masses to show another, and much more conspicuous, effect.

When the rate of contraction varies from place to place, it must result in noticeable strains. An analogy is a tablecloth that has been splashed with a chemical substance. If this is of the kind that causes the fabric to shrink, and if it lands on the cloth in spots, the spots will be areas of shrinking, while the surrounding cloth will not shrink. The result will be that the spots are surrounded by puckers and bulges.

If the cloth is patterned, any lines through and near the spots that were previously straight will have become curved after the chemical has done its work. Suppose that a teacher of geometry had previously drawn lines and triangles on the cloth in order to demonstrate one of Euclid's theorems to his pupils. The lines and triangles will have been distorted by the action of the chemical and will no longer serve as a means of demonstrating Euclidean geometry.

It has been inferred above from Symmetrical Impermanence and general relativity that every heavenly body is analogous to a spot on the tablecloth. It is a region of local shrinking. So the space around it is strained, distorted. Lines in the neighbourhood that would be straight if the body were not there will be curved as a consequence of the extinctions that are occurring in the body. In the region around it Euclidean geometry will not hold; it will be replaced by another kind.

Such is the prediction that is inferred, without any additional hypothesis, from Symmetrical Impermanence and general relativity. Can it be verified? Can one think of an observation or an experiment by which to test it?

This was the question that I put to myself a number of years ago when I was seeking for means of testing Symmetrical Impermanence by observation or experiment. It came as something of a shock at the time that one would be able to observe the effect of the extinction of matter on the geometry of space by the movement of a body free of restraint. If the space was flat, such a body would move with a constant velocity, which means with zero acceleration. But according to relativity theory the body would have a finite acceleration if the space was curved.

Here was the possibility of an experimental test for Symmetrical Impermanence. The hypothesis predicts that a body free from restraint will be accelerated in the vicinity of the earth. The simple experiment of dropping something shows, of course, that it is so. The experiment is cheaper than many to be seen in laboratories and easier to perform. But cost and difficulty are not good measures of the cogency of an experiment. This one would not be more conclusive if it were rarer, more costly or more difficult.

25.3: *The New Theory Developed*
The above remarks have been presented in the form of a justification of Symmetrical Impermanence; and that is what they are. I have shown that the cosmological model inferred from Symmetrical Impermanence without any additional hypothesis is such that the region around every massive body is one in which there is a field of gravitational force.

With a small shift of emphasis and slightly different wording the same remarks would have appeared as a new theory of gravitation. The subject is important enough to make repetition in this form excusable.

In Chapter 21 I pointed out that the word 'mass' may mean three things that can be conceptually distinguished. The names given to them were inert, gravitational and attracting mass. Einstein's general relativity is based on the identity of the first two, both numerically and conceptually. Where there is curvature of space, Einstein pointed out, one can infer that a body possessing inert (and therefore also gravitational) mass is accelerated if it is free from restraint. If the body is near an accumulation of inert mass it is observed to be accelerated. We know this from observation that an accumulation of inert mass causes the space around it to be curved. But this fact is derived from observation only and not from inference. It has not been shown that it is in the nature of inert mass to *cause* curvature. General relativity goes no further than to show that it is in the nature of inert mass to *follow* curvature. Hitherto we have been able to do no more than believe in a vague way that every particle possessing inert mass carries an environment of curved space around with it, but we have not been able to say why.

The new theory that I am presenting here offers an explanation. The particle, I am claiming, does not carry this hypothetical environment about with it. During the particle's continued existence, it has no attracting mass but only inert and gravitational mass. It is at the moment of its extinction that the phenomenon to which the name attracting mass has been given appears. The extinction of the particle is coupled with the simultaneous extinction of some space; there is a local contraction. This becomes manifest as a local curvature of space, the condition shown by Einstein to : define a gravitational field.

Thus gravitation is not the signature tune of matter; it is its swan song.

The local contraction does not remain stationary. One cannot regard it as curvature bound to anything in the sense in which the electric flux around an electron is bound to the electron. As there is nothing to which the gravitational curvature can be bound, it is free in the sense in which the electric flux in a photon of radiation is free. Hence the local curvature that results from the extinction of a particle travels outwards from the site of its source, flattening as it does so. Gravitation occurs in pulses. It is quantized. It has a finite velocity of propagation, though I cannot think of any means of either measuring or inferring the velocity.

The number of pulses of gravitation that emanate per second from any body of which one can measure the attracting mass, m_a is very great. Hence they give the appearance of a continuous field. It is the same with a lump of radium. The radiation is quantized but there are so many disintegrations of atoms to provide the quanta that the radiation gives the impression of continuity. It is the same with light from a lamp. Continuous though it seems, it is really intermittent.

According to the traditional theory of gravitation every particle present contributes its share to the gravitational field. Hence the contribution from each proton or neutron is assumed to be a minute fraction of the total. Though each of these particles is assumed to carry an environment of bound curvature around with it, the curvature around each is regarded as slight.

According to the new theory, on the other hand, the proton or neutron contributes nothing during its continued existence, as has been pointed out already. The contribution comes only from the minute fraction of all those present that happen to be becoming extinct at the moment. The effect of each extinction on curvature at that moment must be correspondingly great. Every extinction, one may say, results in a comparatively powerful jerk to any nearby object in the path of the pulse. The intensity of the pulse, like that of the light from a lamp, diminishes, of course, in con formity to the inverse square law so that the acceleration given to a body by each pulse is less, the further away the body is from the site of the extinction that caused it.

Just as an extinction is surrounded by contracting space, so, according to the new theory, an origin is surrounded by expanding space. In this respect, if in no other, origins and extinctions are distinguished only by sign. So the space around an origin must also witness the passage of a pulse, or wave, of curvature. This curvature must have the opposite sign to that of a gravitational pulse. Instead of being a wave of contracting space it is a wave of expanding space. It can aptly be called a wave of anti-gravitation. Its effect must be to accelerate any inert mass over which it passes away from the site where the origin occurred.

It has been shown that the rate of origins must everywhere be constant per unit volume. Although each origin is a centre of anti-gravitation, the random and approximately uniform distribution of these centres throughout space prevents the anti-gravitation from becoming observable. Pulses of anti-gravitation are passing everywhere in different directions and usually cancel each other's effects.

According to the new theory things do, nevertheless, fall away from the site of every origin in a very tenuous gas. Origins are sources of dispersal, just as extinctions are sources of concentration.

A region, moreover, in which the mass density is below the equilibrium value is one from which more pulses of anti-gravitation emanate than pulses of gravitation. Most extra-galactic space consists of such regions and they must cause inert mass to move out of them. The things that fall towards galaxies are, one might say, being pushed there out of almost empty space as well as being pulled by the galaxies.

So much for a brief outline of the new theory. It will be shown in Chapter 26 that it has sufficient explanatory power to answer the questions raised in Chapters 21 and 22, and it will be shown in Chapters 27 and 28 that it also helps to explain the formation of stars and their rotation.

Chapter 26

Answers to Questions About Gravitation

Particularly in Chapter 22, but also elsewhere, a number of questions about gravitation have been raised to which the traditional theory has no answers. But for most of these questions answers are provided by the new theory and these will be presented here. Before this is done, however, a com- parison of the statements about gravitation that are implicit in each of the theories will be helpful. It is given in parallel columns below:

Traditional Theory	*New Theory*
I. The cause of a gravitational field is the presence at its source of an accumulation of inert mass.	The cause of a gravitational field is the extinction at its source of elementary components of the material universe.
2. Every particle at the source of a gravitational field contributes a share to the field strength, and this contribution is proportional to the inert mass of the particle.	Only those components contribute to the strength of the gravitational field that are becoming extinct at the moment.
3. The hypothesis on which the theory is based is the *ad hoc* one that every body with inert mass also has attracting mass.	The hypothesis on which the theory is based is the Principle of Minimum Assumption. From it the Symmetrical Impermanence of matter is inferred and from this, in turn, the new theory.

4. The immediate basis of the theory, namely the identity of inert and attracting mass, has no other manifestations	The immediate basis of the theory, namely Symmetrical Impermanence, has many other manifestations. Among them are the occurrence of interstellar gas, the expansion of space, the occurrence of discrete nebulae, their detailed structure. The more remote basis of the theory, namely the Principle of Minimum Assumption, has vastly more manifestations and is the widest of the generalizations with which physicists work.
5. The observed constant ratio between the strength of the gravitational field and the quantity of inert mass at its source is the consequence of a specific law	The observed constant ratio is not the consequence of any specific law, and such laws are precluded by the Principle of Minimum Assumption. The constant ratio is the consequence of a statistical law, which says that in a sufficiently large sample of matter the rate of extinctions per unit volume must be nearly constant.
6. The gravitational field is not the consequence of any change. It depends on what is, not on what happens.	The gravitational field is the consequence of a change, namely that from existence to non-existence of an elementary component. Thus it does depend on what happens.
7. Gravitation is unique among physical phenomena in being continuous	Like a beam of light, which consists of separate photons, a gravitational field consists of separate pulses. It is thus quantized and, in this sense, intermittent.
8. As the source of a gravitational field does not undergo any change in producing the field, it does not lose any energy in doing so.	The source of a gravitational field loses the energy that is represented by the mass of the elementary components that are becoming extinct.

9. As the source of a gravitational field is not losing energy in producing the field, gravitation is, unlike light, sound and all other phenomena, not the consequence of the transmission of anything from one place to another.

Like light, sound and all other phenomena, a gravitational effect only occurs when something is transmitted from one place to another.

10. Being unassociated with change of any sort, a gravitational field cannot consist of waves.

The gravitational field can and does consist of waves of curvature in space. They are set up whenever a particle becomes extinct.

Answers to the various questions about gravitation that have hitherto proved puzzling are largely (and I hope wholly) implicit in the above list of statements. Those that have been asked in Chapter 22 are answered below in the order in which they occur.

Question: Can gravitation be brought into association with the main body of science?

Answer: The reason why this could not be done in the past was that the traditional theory is based on an *ad hoc* hypothesis. This was that a specific law, applicable only to objects with inert and gravitational mass, requires them also to have something quite different, namely attracting mass.

It has been explained in Chapter I that *ad hoc* hypotheses and unified theories are incompatible and a brief recapitulation will be helpful here. Before Newton there was the ad hoc hypothesis that a specific law, applicable only to planets, required these to move in elliptical orbits. Had Newton been content with this he would not have achieved the unification of all mechanics that has done so much for science. He would have left many questions about planets, and other things, unanswered that he clarified when he showed that many apparently isolated phenomena are really special manifestations of general principles. Once the elliptical orbit of planets was no longer explained by an *ad hoc* hypothesis it found its place in the main body of science.

It is similarly pointed out in the third of the above statements that, according to the new theory, gravitation is one of many manifestations of a general principle, namely Symmetrical Impermanence. This alone would suffice to bring gravitation within the main body of science. But the unification goes even further.

The Principle of the Symmetrical Impermanence of Matter is, in turn, not based on an *ad hoc* hypothesis, but is an inference from an even more general principle. This is the Principle of Minimum Assumption. Sym- metrical Impermanence appears as but one of very many manifestations of this. A more general principle could hardly be thought of. It both precludes all *ad hoc* hypotheses and provides a roof over a complete and self-consist- ent structure of scientific thought in physics in which gravitation finds a place and bears a logical relation to all other parts.

Question: Why do inert and attracting mass always occur in association?

Answer: They do not. Matter has no gravitational effect during its continued existence. Each particle exercises this effect only at the moment of its extinction.

From statistical considerations one should, of course, expect to observe strict proportionality between the quantity of inert mass present in a body and the gravitational field that surrounds it. For the extinctions that produce the field must occur at a rate proportional to the amount of mass. But we have been misled in the past by this proportionality into the belief that every atom is contributing to the field all the time.

It is interesting to note that a similar hypothesis never seems to have been seriously entertained about radio-activity. The notion that every atom in a lump of radium contributes to the radiation all the while has had little, if any, support. We should today smile at anyone who insisted that it must be so. We know that the activity occurs only at the moments when an atom is disintegrating. We do not regard radio-activity as the signature tune of each atom of radium but as its swan song. Yet the metaphor of signature tune is still commonly regarded as the only appropriate one for gravitation.

Question: Why does gravitation seem to contradict the Principle of Stabilization, according to which every process in a self-contained system tends to reduce its cause?

Answer: It only seems to do so and this because we have hitherto wrongly assumed the system in which a gravitational field occurs to be self-contained. Such a system is one in which matter neither originates nor becomes extinct. That in a gravitational system processes tend to increase their causes does not prove, therefore, that the principle does not hold universally. What it provides evidence for is that a system in which there is gravitation is not a self-contained one.

Question: Why do gravitational fields always have the same sign?

Answer: Because they have only been observed where the mass density enormously exceeds the equilibrium value. The rate of extinctions per unit volume therefore enormously exceeds the rate of origins. As the sign of the field depends on which rate exceeds the other, one must expect the sign in the vicinity of massive bodies always to be the same. But the observed constancy of the sign is the result of statistical laws and not of a specific law.

A region where the rate of origins exceeds the rate of extinctions per unit volume must be surrounded by a field of the opposite sign. In such a region the mass density is below the equilibrium value.

Question: Why is there a limited number of constant ratios between unit electric charge and the inert mass of an elementary particle?

Answer: This question is not about gravitation at all. It is about inertia and electric charge. One must not expect the new theory to provide an answer and it cannot do so. But, nevertheless, the question has some relevance here. For the conclusion that gravitation is quantized may give a new urgency to a search for a principle by which all the indivisible physical quantities are co-related.

Question: Why is the gravitational field attributed to an elementary particle vastly weaker than the electrostatic one?

Answer: It is not necessarily so. Let t be the time during which an extinction is being effective and let $1 / n$ be the fraction of the particles present at the source of the field that are becoming extinct during this time. Of course, n is a very large number. The contribution to the field made by a particle is then n times the contribution attributed to it by the traditional theory and may be of similar magnitude to the electrostatic field around an electron.

There must be a quantitative relation between the duration of a pulse of gravitation, t, the number n, and the half-life of matter. If two of these quantities were known, the third could be calculated. But at present none of them are known. It will be worthwhile to look for pointers to the value of these three quantities.

Question: Why is gravitation uncontrollable?

Answer: Because the events that cause it, namely extinctions, are uncontrollable. Their uncontrollability is, indeed, intrinsic and it would be idle to attempt to overcome it. According to Symmetrical Impermanence, origins and extinctions are not associated with anything in the existing state of affairs. If they were, they would not be true origins and extinctions, but only conversions. Hence anything that one could do to influence the existing state of affairs could not have any effect on origins and extinctions. It follows that a

184

given inert mass must always be surrounded by the same constant gravitational field whatever be done to it or to anything else.

Question: Why is gravitation the only phenomenon that does not seem to be the consequence of any change?

Answer: It is not. Like every other known phenomenon gravitation is the consequence of a change and the change is the most radical that can be conceived. It is the change between existence and non-existence of an elementary component.

Question: How can one reconcile the conceptual difficulty of action at a distance with what is known about gravitation?

Answer: The distance between the place where the extinction occurs and the place where a particle is accelerated is spanned by the movement of a wave of curvature. One can think of this as a pulse of gravitation. Like a letter, a projectile or a photon of radiation, it is a carrier of something and transfers a change in the geometry of space from one place to another. There is, therefore, no action at a distance. The action of gravitation, like that of light, is effected by the transmission of a signal.

I should hesitate, however, to suggest that the pulse of gravitation can helpfully be regarded as a carrier of energy. I am strongly inclined to deny it. It is true that each particle over which the pulse passes is accelerated to a velocity, v. We can speak of the particle as having acquired the kinetic energy, $1/2\ mv^2$. But we may not speak of this as having been acquired at the expense of the energy in the pulse; for the energy in the pulse does not seem to be changed by the encounter. If it were changed, there would be such a thing as screening against gravitation and we know that there is not.

It must not be forgotten that gravitational fields can only occur because the system is not self-contained. Hence the Principle of Conservation of Energy is not applicable and one must be very cautious about drawing inferences from this principle.

Question: There is, further, a question that occurred in Chapter 7 Why do we not observe an infinite intensity of gravitation here and now?

Answer: It will be remembered that we should expect to observe an infinite intensity of radiation if it were not for the fact that the universe is expanding. A consequence of this is that there is an optical horizon. No radiation can reach us from places beyond this.

According to the new theory there must also be a gravitational horizon. There must be a critical distance beyond which any two objects are being separated at a rate greater than the velocity of propagation of the gravitational waves.

As the velocity of the waves is not known, this distance is not known either. Perhaps a method for discovering its value will be found some day.

Chapter 27

The Formation of Stars

27.1: *The Problem*

The conclusion was reached in Part Three that a cosmological model based on Symmetrical Impermanence would contain spiral nebulae. In this respect it would resemble actuality. But it did not seem to do so in another respect, for no reason could be found why the nebulae should contain discrete stars. It seemed that they must consist only of a more or less diffuse gas.

For many years this conclusion led me to doubt the validity of Symmetrical Impermanence. It seemed to lead to a false model. But it will be shown below that what was at fault was not Symmetrical Impermanence but the traditional theory of gravitation. If one adopts the new theory that has been presented in Chapter 25, one arrives at a model in which the nebulae do contain discrete stars.

The difficulty of arriving at the correct model with the traditional theory is, it will be remembered, that, according to this theory, the gravitational pull on particles of gas in a cosmic cloud would always be towards the centre of gravity of the whole cloud. But stars would not form if movement were exclusively towards the common centre. The result would be one single, compact mass. For stars to separate out there must be movement towards local centres; some particles must move outwards, away from the centre of gravity of the whole cloud and towards the centres of the incipient stars. So star formation would require the setting up of parochial fields of force strong enough to compete successfully with the gravitational field common to the whole cloud.

Of course one could save the traditional theory of gravitation by inventing additional hypotheses. Perhaps, one might say, some other kind of force operates to cause stars to condense out of a cloud of gas. Perhaps the stars have not condensed out of a cloud of gas at all but have beer assembled by some quite different and unknown process. Perhaps the Cosmic Statute Book contains a law to say that particles shall congregate to form stars and they obediently do so. But 'perhaps' as the basis of an explanation is a word that should be avoided in physics whenever possible, An hypothesis that cannot be inferred from the Principle of Minimum Assumption has, I should like to insist once again, no place in physics. But it will be shown below that no *ad hoc* hypothesis is needed to explain the formation of stars.

27.2: *In a Tenuous Cloud Gravitational Pulses are Intermittent*
The explanation of star formation lies in the simple fact that the gravitational force is, like light, intermittent. Around any appreciable concentration of matter the discontinuity is no more observable than the discontinuity in a beam of radiation, which, nevertheless, consists of a bundle of discrete photons. But when the quantity of mass in a given place is quite small, there must be a significant interval of time between successive gravitational pulses.

In a gas so diffuse that scattered atoms are well separated from each other the sites for extinctions must be few and far between. According to the new theory of gravitation, it has to be remembered, these isolated atoms do not exercise any gravitational effect at all so long as they continue to exist. It is only when, here and there, an atom becomes extinct that a gravitational pulse is emitted. The effect of such intermittent extinctions is very different from a steady pull in any one direction.

After there has been an extinction in one particular place there may not be another in the vicinity for an appreciable time. The next nearest extinction may be a great distance away if the gas is very diffuse. Such infrequent extinctions will impart spasmodic jerks to all matter that comes under their influence. During the intervals of time between them there will be no forces at all except those that arise from still more remote extinctions; and as the pulses diminish in intensity according to the inverse square law the more remote sources of quanta of gravitation will have a relatively feeble effect.

In such conditions it is rather meaningless to speak of a centre of gravity for the whole cloud. There is a geometric centre, but the gravitational pull is by no means always and everywhere directed towards it. The pulls are random; they arise from scattered sources and operate in all directions. At most one can say that there are, on the average, more pulls in the direction where there is most mass, namely towards the geometric centre, than away from it.

27.3: *Incipient Concentrations*
Let us consider in the light of the above remarks what must happen to particles in the cloud that has begun to form around an astronomical summit. The density of such a cloud, it has to be remembered, is by terrestrial standards very low. A gramme occupies a vast region.

Let us first consider extinctions. These produce pulses of gravitation. So long as such a passing pulse of gravitation lasts, all particles in the cloud that are reached by it experience an acceleration towards the site. When the pulse is over there is no more acceleration, but the particles have acquired a finite velocity and are therefore converging from all directions on to the site. Origins have an opposite effect. As these are distributed almost uniformly in space, the dispersals occasioned by them are on the average as much away from as towards any particular direction. They must create a little turbulence in the gas, but do not have any lasting effect on the distribution of its particles. They may, however, break up concentrations that are beginning to form. For after an extinction has caused particles to converge on to the site, a subsequent origin in the same place will send the particles back to where they came from. This will prevent every extinction from leading to a lasting concentration. But one can infer that some of the incipient concentrations will not be dispersed before they have established themselves. The establishing of at least some follows from statistical considerations.

The probability of an extinction anywhere is directly proportional to the gas density there. When, therefore, the particles have begun to crowd together around the site of a recent extinction, it becomes a little more probable that another extinction will occur near the previous one. When this does happen the converging particles will receive an additional acceleration in roughly the same direction in which they are moving already. They will then not be so easily scattered as before.

The second extinction will cause the crowd to thicken and make a third extinction there yet more probable. If this occurs, the crowd of particles will be rendered still more dispersal-proof. So it must go on. After a concentration has got well underway it must steadily increase.

The process just described is of the kind that can be achieved in engineering with the help of devices that provide what is called 'positive feedback'. But no such devices are needed to maintain the process of forming concentrations in extragalactic space. The laws of probability suffice to explain what happens. When the process has continued for long enough, the result is a very big and massive concentration. We call it a star. Here, incidentally, is an example of a process that tends not to a stable equilibrium, as with negative feedback, but to increase its cause.

27.4: *Incipient Stars have an Irregular Structure*
Probability considerations lead one to expect an increase in the number of extinctions around the geometric centre of an incipient star during the very earliest stage of its formation; but these extinctions may be near to or far from the geometric centre. Each will form its own incipient concentration so that the structure of the incipient star must be far from homogeneous. Within this structure there must be numerous parochial concentrations. But as the star becomes more dense and pulses of gravitation within it more frequent, conditions must gradually change so that the general effect is as though all the pulses emanated from the common centre. The intervals between pulses will become so short that particles will never have time to move far towards the parochial centre before they are accelerated away from it towards the centre of gravity of the whole star.

Before this happens the incipient star must have a quite irregular shape. But

189

as extinctions within it become numerous enough to be practically continuous, it will be gradually pulled together into the shape of a homogeneous sphere.

27.5. *Double and Multiple Stars*
As the thickening of the gas extends to a substantial distance from the incipient star, there must be many occasions when an extinction at a substantial distance forms the beginning of a second star. If this happens while pulses from the original incipient star are still significantly intermittent, the second star will be able to form, competing for hydrogen successfully with its slightly more developed neighbour. More than one star may thus form at a certain distance from the first star. The consequence will be one of the double or multiple stars that are rather numerous.

It should be noted that the components of double or multiple stars must, according to the new theory, all begin at about the same time. If one of them had become well-established it would be so massive that it would attract all surrounding particles to itself; a later incipient concentration would not stand a chance of surviving.

The component stars must also be of approximately the same size during their period of growth. For if any one of them were much less massive than its neighbours it would be absorbed by them. Small differences in mass need not, however, prevent double or multiple stars from growing side by side during the time while they remain very tenuous.

Later, however, when the difference between gain by capture and loss by extinction becomes very narrow a small reduction in income from capture can lead to a change from growing to dwindling. One should therefore expect the two components of a double star to have unequal later histories. When both have reached the size at which income and loss balance one should expect the slightly larger one to maintain, even somewhat increase, its mass while the smaller one would lose mass at an ever increasing rate. In other words, neighbouring stars of the size at which income and loss nearly balance are in competition for new matter and the smaller ones lose the battle. The consequence of this will be discussed in Appendix D.

27.6: *At What Gas Density do Stars Form?*
The gas density at which stars can form would seem to be a quantity of some significance in cosmology. Knowledge of it might help us to estimate the stage in the evolution of a spiral nebula at which star formation occurs. But at present it is only possible to say that there must be an upper and a lower limit.

That there is a lower limit is obvious, for no stars can form where the density is zero. The density must be at least such that the conditions for the effective crowding together of particles are met. These conditions are that the particles are able to acquire a significant average velocity from one single quantum of gravitation and that there is a significant number of such particles. But the velocity cannot be significant if the nearest particles to the site of an extinction are a long way off. The pulse of gravitation can only produce such a velocity if it has not become too weak; and its effect diminishes with the square of distance.

Hence the pulse can only significantly affect the movement of particles within a limited range. If the number of particles within this range is small, the effect on the local density will be negligible.

It has to be remembered that the scattering effect produced by origins is independent of density. There are as many quanta of anti-gravitation per unit volume in a high vacuum as in a dense medium. But a large number of particles moving at a high velocity are not so easily dispersed by a single quantum of anti-gravitation as a small number moving slowly. For this reason the chance of becoming dispersal-proof increases up to some limit as the density increases.

The upper density limit is reached when particles are so close together that extinctions in the near vicinity follow each other in close succession, The particles will then be accelerated towards the region of the greatest number of extinctions, which means towards the geometric centre of the system. The effect of the whole mass predominates at a certain density over the parochial effect of nearby particles.

Once star formation has become established in a cloud it must become increasingly difficult for new stars to begin. The competition of the established ones must be too great. One should therefore expect most of the stars in the core of a spiral nebula to begin at about the same time. The possibility, however, of the occasional formation of new stars on astronomical summits within a spiral nebula ought not to be disregarded.

The history of the stars in the spiral arms may be a little different. The density at which the original population of stars begins must be very low; so one should not preclude the possibility that stars may begin to form while the spokes of the cloud are still lying on the astronomical shoulders and long before these have poured into the rotating core to form the spiral arms. Indeed, there are considerations that point in this direction.

After the pouring has occurred there must be considerable turbulence, The spokes have entered the core from an enormous height and must stir things up considerably as they splash into the core. I have already suggested in Chapter 20 that this turbulence ought to be picked up by a radio-telescope. Now it is not easy to believe that the little crowds of particles that are assembled around the site of a recent extinction could ever survive such a bufieting. From this I am inclined to conclude that the crowds have become fairly massive in the calm atmosphere of the spokes as these rest sluggishly on the astronomical shoulders. They must be large and compact enough to be the sites of concentrated extinctions. Only so can they be dispersal-proof, even from the violence of the agitated gas in which they find themselves after the pouring process.

On the other hand, the concentrations can hardly yet approach the massiveness of stars before the pouring process. If they did, they would be only slightly delayed in their fall by impact with the gas of the core. They would all fall deep into the interior and one should not expect to observe any stars in the spiral arms.

The tangential effect of the gas in the core would, moreover, be no greater than the radial effect. Not only would massive stars fail to stay near the surface, if they did they would not be entrained.

Thus we are led to the conclusion that star formation probably begins in the central core at an early date and later also in the spokes, far out beyond the limits of the future nebula. Formation of these stars is completed in the spiral arms.

Chapter 28

Why do Stars and Galaxies Rotate?

28.1: *The Problem*

It is known that stars rotate about an axis. So do galaxies. Why? The answer must conform to the principle of conservation of angular momentum and it is this that makes it difficult to find an answer. The angular momentum of a system is the product of its moment of inertia and its angular velocity. The principle of conservation of angular momentum states that the value of this product is not changed by any forces that act only between the component parts of the system. To change the angular momentum it is necessary to apply a couple from outside.

So long as no external couple is applied, any change in the moment of inertia of the system is accompanied by a corresponding change in its angular velocity. Provided the mass remains constant, the moment of inertia is proportional to the radius of gyration. In a system to which no couple is applied, and of which the mass is constant, the radius of gyration varies, therefore, in inverse ratio to the angular velocity. This is why contraction of a star under the influence of its own weight is accompanied by an increase in its rate of spin.

The angular momentum of a star, such as the sun, is considerable. We know from the principle of conservation of angular momentum that this large quantity has been imparted to the sun at some time during its past history and that a couple must have been applied to do this. What was the couple? How did it act? When was it applied?

The same questions can also be asked concerning the enormous angular momentum of the galaxies and if they can be answered for the one the same kind of answer should hold, broadly speaking, for the other. So it will be convenient to discuss the problem mainly for stars. It is more hopeful to look for the couple during the star's earliest history than later, so attention will be directed towards the incipient star.

One might for a moment be tempted to think that a non-uniform distribution of the substance of the incipient star would suffice to generate angular momentum during the process of contraction from a tenuous gas into the compact star. For any asymmetry would cause the average movement of the particles that were falling inward to be off-centre. Their force of impact with

193

other particles would produce a couple and this would lead to a swirl about the centre of gravity of the star. An accumulation of such swirls would amount to rotation provided they were all about the same axis and in the same sense of rotation.

But they would not all be like that. There would be swirls in all directions, left hand and right hand ones. These would collide and lead only to turbulence. A swirl in one direction could never preponderate over the others, even to a small extent, however irregular the shape of the incipient star was, so long as movement of particles was random. If the component parts are at rest relative to each other when the contraction begins, a mere process of contraction cannot generate the least spin of the star about its axis.

This follows quite simply from the principle of conservation of angular momentum. The forces that lead to contraction are forces between the component parts of the incipient star and the principle asserts that these cannot generate any angular momentum. This was zero when the contraction began and so it must remain zero unless an external couple changes it

It might, of course, be argued that the component parts were never at rest relative to each other, and this is probably true. One may expect some thermal agitation, some collisions between molecules. But a moment's reflection shows that these cannot constitute angular momentum.

The movement of a particle has two components, one along a line that joins it to the centre of gravity of the incipient star; it is a radial component. The other is at right-angles to this line, a tangential component. Only the tangential components could contribute to angular momentum and to do this there would have to be more components in one plane and direction of movement than in the others. But with random movement of particles in thermal agitation this does not happen. For thermal agitation means random movement of a very large number of very small particles. On probability considerations it is easy to show that the vector sum of all the momenta must be zero to a very close approximation.

Can one then justify the assumption that large masses of gas move in solid phalanx in a particular direction? Suppose there were a limited number of such phalanxes and the movements within each were not random. Each phalanx would have a tangential component of its velocity and momentum. If the vector sum of the momenta were taken, there would be a residual vector in a particular direction. With a random distribution of phalanxes it would be small if there were many of them, but not negligible if there were not many. Such a residual vector would amount to angular momentum. But the difficulty is in finding a reason for the phalanxes. Why should a large number of particles move together in the same direction? Where is one to look for a co-ordinating force, for the guiding principle that would preclude random movement of the individual particles ? By this line of reasoning angular momentum would be accounted for only by the assumption of something that acted like a cosmic drill sergeant. But such an assumption would be at least equally difficult to explain. There does not seem to be any tenable alternative to attributing the rotation of stars and galaxies to an initiating couple, and it is very difficult indeed to account for this or to understand how it can act. Facile explanations have been provided for the rotation of stars, it is true, and sometimes with the backing of high authority. But they are based on flimsy reasoning, which collapses at the first breath of criticism. Let us consider the problem in the appropriate terms, namely those of simple mechanics.

28.2: *The Conditions needed to cause Continued Rotation*
For continued rotation to occur in a system it is axiomatic that the following three conditions must be met:

(a) An external couple must be applied to the system.

(b) The system must have such asymmetry as enables the couple to act on it. Thus the slot in the head of a screw allows a screw-driver to act on the screw. The asymmetrical structure of a compass needle, with N at one end and S at the other, enables a bar magnet brought close to the compass needle to exert a rotating couple. In consequence the needle turns until one pole points towards the bar magnet. A crank provides a similar asymmetry and enables the force exerted by steam on the piston of an engine to impart a couple to the flywheel that is rotated.

(c) The external couple must either rotate with the system to which it is applied or cease to operate after rotation has been initiated. The first requirement is met by a screwdriver. This rotates with the screw. The second requirement would be provided if a bar magnet brought near to a compass needle were quickly withdrawn after rotation had been initiated. Provided the needle were shielded from the earth's magnetic field it would continue by virtue of its inertia. But when the bar magnet is not withdrawn, the needle comes to rest in a position in which it points towards the magnet This is a position of minimum potential energy, of equilibrium. A technical name for it is 'dead centre'. For rotation to be continuous the applied force must be such that it fails to hold the rotating system at dead center.

In order to explain the rotation of stars and galaxies one must find circumstances that meet these conditions.

28.3: *Accounting for Asymmetry*
To meet the first condition the incipient star must be subjected to preponderating force from one particular direction. It would not rotate if subjected to forces from many directions of which the vector sum wasl zero.

The biggest unbalanced force would come from a neighbouring star. In particular, two components of an incipient double star must exert a gravitational force on each other. This must greatly exceed the gravitational force exerted by more distant stars.

To meet the second condition the incipient star must itself be of irregular shape. But it has been shown in Chapter 27 that the incipient star would be of such a shape. It would be a collection of incipient concentrations with an irregular distribution. Only at a later stage of development would the star be pulled together by its own gravitational field into a homogeneous sphere. For dynamic purposes the structure of the star could be resolved into a dumb-bell shape. The two concentrations that were equidistant from the geometric centre would be the equivalent of two halves of the star's mass.

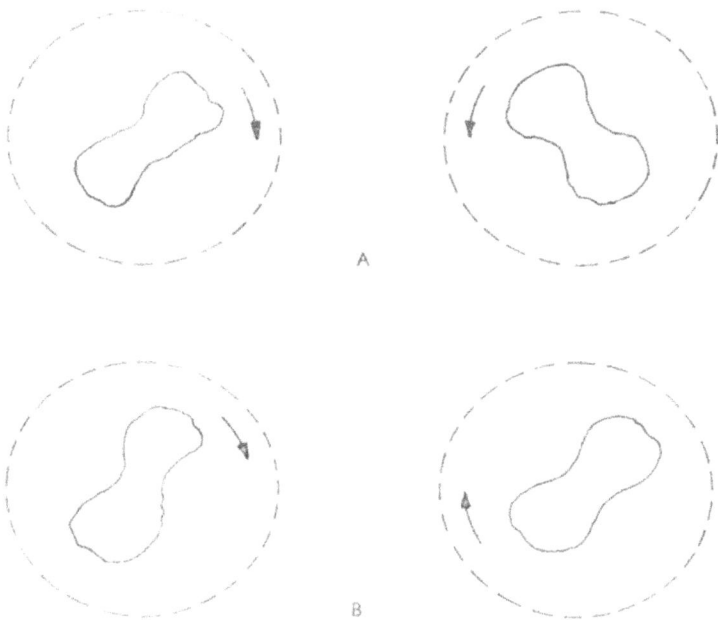

Fig. 5. Rotation of incipient concentrations under the mutual effects of asymmetrical structure, intermittent gravitational pulses and gradual dis- appearance of asymmetry under contraction

Two neighbouring stars, when represented by their equivalent dumbbells, could have, at a given moment, one of the two relative positions shown in Fig.

5 at A and B. The nearer mass would be drawn towards the neighbouring star more strongly than the further one. The consequence would be that gas particles in the nearer mass would acquire a greater velocity towards the neighbouring star than those in the further mass. If the dumb-bells were rigid structures, they would rotate until they lay on a common axis, provided the attracting force were continuous. In fact they are not rigid structures. Each gas particle acquires a velocity towards the neighbouring star as well as a velocity towards the geometrical centre. This leads to a component of its movement at right angles to the above-mentioned radius, r. But the component of particles in the concentration nearest to the neighbouring star predominates. If the component is maintained, collisions between gas particles will lead to a net movement around the geometrical centre in the direction shown by the arrows.

28.4: *How Dead Centre is Passed*
It has been said above that the dumb-bells would align themselves along a common axis if the attracting force were continuous; and it would be continuous according to the traditional theory of gravitation. But it is not continuous according to the new theory. While the incipient stars are still very tenuous, the pulses of gravitation between them are intermittent.

This fact makes it possible for the irregular rotating system to pass dead centre. True, the pulses do not diminish in intensity and they increase in frequency as each of the stars grows. So one might think that the couple that initiates rotation would increase and hold the dumb-bells all the more firmly to dead centre when it was reached. But there are two ways of reducing a couple.

A couple is the product of a force and the projected length of a lever arm. The couple may be eliminated by removing the force, as happens when one quickly withdraws the bar magnet from the vicinity of a compass needle. But the couple may also be eliminated by eliminating the asymmetry of the system. This would happen to the compass needle if its magnetism were to cease.

The lever arm of the incipient star is represented by the average distance between the parts represented by the halves of a dumb-bell and the geometric centre. As the star contracts this distance decreases and therewith, the capacity of an external force to produce a couple decreases too.

Thus each successive pulse coming from the neighbouring star finds less irregularity to 'get hold of as it were. It is during the intervals between pulses that the star is enabled to pull itself into homogeneity. By the time, the position that would be dead centre is reached there is a less pronounced dead centre; for the star has become more symmetrical.

Galaxies do not have companions capable of exerting couples. But it will be
197

remembered that, during the first stage of growth, the core is surrounded by the enormous spokes. These are irregularly spaced and the core has developed around an astronomical summit of irregular shape.' The requisite asymmetry is there, both external and internal, to initiate rotation of the core.

PART V

APPENDICES

Appendix A

What Further Steps Towards Unification?

A.1: *Irreducible Concepts*

This book represents but one step in the direction towards the unification of physical science. Many previous steps have been taken by scientists and many more will have to betaken before unification is complete. Perhaps that never will be, but it is desirable at least to make the effort. Past success itself justifies it, for already the discovery of ever wider generalizations in physics and our increasing knowledge of the relation between apparently isolated phenomena have greatly diminished the number of irreducibles with which physicists have to work. Those that remain can be placed in three classes, namely:

(a) Irreducible physical laws and principles,
(b) Irreducible physical constants,
(c) Irreducible constituents of the material universe.

The members of each class are today still too numerous to make complete unification look very near, but not so numerous as to make it look hope- less. In this first appendix I want to do no more than define the problem as I see it. Detailed discussion would not be in place here. I shall therefore not attempt to elaborate such suggestions as I am putting forward but merely express the hope that they may encourage further research.

A.2: *Tautologies*

Some of those generalizations that are called laws or principles in physics are tautologies: their truth is implied by their verbal formulation. These form a big field of study and are of great importance in scientific method, but again this is not the proper occasion for their detailed discussion. It must suffice to note three facts about them.

(Here I am using the word 'tautology' to designate a statement about a concept that is implicit in the definition of the concept. Such a statement repeats in different words (or mathematical symbols) what is said when the concept is defined. As technical terms in philosophy and logic have never been standardized and are used with different meaning by different people there must be some who would prefer a different word to 'tautology'. But that need not affect what I have to say here.)

Firstly, far from being superfluous, tautologies serve a most useful purpose in physics and are more abundant than is often supposed. Mathematical statements belong to this category. The multiplication table, for instance, is implicit in the definitions of number and of the process of multiplication. In a similar way it can be seen that some of those scientific laws that can be expressed in algebraic form are tautologies. They provide a means of expressing in letter symbols what is implicit in the concepts that are being handled.

Secondly, tautologies, like everything true by definition, are not only sometimes true or only approximately true. When they are formulated correctly they are always true and necessarily true. This characteristic helps with the detection of tautologies. One need not look for them among those looser statements that may be subject to exceptions or no more than first approximations. Nor need one look for them among experimentally falsifiable statements.

Thirdly, and consequently, a tautology does not have to be verified by experiment or observation. It can be proved with the help of pencil and paper guided by logic. This holds even though experimental confirmation of a tautology may also be possible. The multiplication table can be verified by experiments with an abacus or by counting oranges, but its validity does not depend on such procedure. This characteristic provides further help with the detection of tautologies. Sometimes a law has been discovered by experiment and its true nature is not immediately obvious. But it may afterwards be satisfactorily 'explained' with the help of pencil and paper. The explanation then proves that the law could have been discovered without the experiment, that it is implicit in the definition of the concepts that are embodied in its formulation, that, in short, it is a tautology.

One part of the work of unifying physical science is the detection of tautologies.

A.3: *Laws that are Implicit in More Comprehensive Ones*
Another part of unification is concerned with those laws that are not recognized as tautologies. Their number diminishes whenever it is found that a law is implicit in more comprehensive ones. It is with this part of the work of unification that the present book has been largely concerned. If its reasoning is sound, several laws have been taken out of the category of irreducible laws and placed in the category of the deducible ones. Among them are the law that requires nebulae to have spiral arms and the one that requires every large inert mass to be the source of a gravitational field. I venture to suggest that the method adopted here could, with advantage to the progress of physical science, be recognized more generally and followed more systematically than it has been in the past.

A.4: *The Reduction of Physical Constants*
The irreducible physical constants with which we are still left today include the velocity of light, the quantum of action, and the electric charges that occur respectively on the proton and the electron. What has been said here in previous chapters suggests that the rate of origin of matter and its half-life should be added to the list. But when all necessary additions have been made, the list is today not a very long one.

The unification of physics requires that a relation be found between the values in the list so that one would be able to say that these values were inter-dependent. If it were so, one could infer one of the values from others. Eddington attempted the task, though apparently without success. There were some at the time who criticized him, oddly enough, not because they had proof that he failed but because they thought he ought never to have tried.

A.5: *Establishing a Relationship between Different Constituents of the Material Universe*
The irreducible constituents of the material universe include an inconveniently large number of so-called elementary particles, the proton, the electron, the antiproton, the positron, many mesons, the neutron, the neutrino. They also include radiation, energy, space and time. The list would have been longer half-a-century ago. It would, for instance, have included ninety-two distinct chemical atoms. Still earlier it would have included all those substances that have been since reduced to their chemical constituents.

Previous reductions have sometimes occurred when it was found that what had been thought of as elementary was really composite. It was so with the reduction of chemical substances to chemical elements and of the elements, in turn, to protons, neutrons and electrons.

Previous reductions also sometimes occurred when it was found that one apparently irreducible constituent could be converted into another. Both then appeared as different manifestations of the same single con- stituent. Ice and water are a simple illustration. These two commodities are reduced to one by substituting for both the single symbol H_2O.

The process can be called unification by substitution. It was particularly fruitful when the concept energy was introduced. This was found to be a commodity that may occur in many different forms. One of them is called work, another heat; in one form it is defined by the symbol $\frac{1}{2}mv^2$, in another it is stored in the electrostatic field and takes the form $\frac{1}{2}CV^2$. Einstein discovered that it may also occur in the form of inert mass.

The unifying power of this discovery was very great. For it showed that mass, which we think of as 'quite real', and energy, which we tend to think of as conceptual only, are both manifestations of one basic reality.

Yet another step towards the unification of physics has often been the complete elimination from this science of some constituent of reality that was previously included in it. Nature's horror of a vacuum is an example. Feelings are today known to have no place in physics but to belong to other disciplines.

Sensations are now in the same category. The sensation of a given colour, the smell and taste of things, aesthetic and ethical values, every- thing in short that depends on the experience of a particular observer, has been excluded from a unified physics.

Constituents that have been postulated for the convenience of the imagination have gone the same way. The reason for their elimination has been that they depend on the observer in his capacity as thinker or image-maker. For this reason Newton's space has been eliminated from the list of constituents. A space of which it can be said that no place differs in any way from any other place is purely conceptual and plays no part in physics. It has been replaced by Einstein's space, which is, as I have suggested elsewhere in these pages, no more than a synonym for environment.

In the same way absolute time has had to go. The notion of undifferentiated time as a background for events has proved as meaningless in physics as the notion of featureless space as a container of things. Einstein's time, I am inclined to think, is no more than a synonym for distance divided by velocity. Thereby the concept, time, has been stripped of its subjective implications and quite a useful step taken towards unification.

A.6: *Brief Summary*
Above, five methods have been discussed in the briefest of words by which the number of irreducibles that occur in physics may be caused to decrease. For convenience let me condense even more and enumerate them in the form of a list. The five methods are:

1. Reduction by finding that something is a tautology;
2. Reduction by finding that something is implicit in some other more comprehensive thing;
3. Reduction by finding that a variety of distinct things are all composed of a small number of elementary components;
4. Reduction by finding that two or more distinct things can convert into each other;
5. Reduction by finding that something does not belong to the physicist's world.

I should not like to venture a guess as to which of these methods is likely to prove the most rewarding. They will probably all have been used lavishly before complete unification has been achieved.

Appendix B

Can a Galaxy Acquire an Infinite Mass?

B.l: *Relative Growth of Neighbouring Galaxies*
Let us consider the boundary around a galaxy that has been called the reversal zone in Chapter 10. This, it will be remembered, is a surface that entirely surrounds the galaxy and at which the potential gradient is zero. New matter that originates within the reversal zone falls towards the galaxy enclosed by it. New matter that originates beyond this zone falls towards a neighbouring galaxy.

The position of a reversal zone anywhere is a complicated function of the distribution of masses within a volume large enough to comprise many galaxies. But let us, as a first approach, consider two neighbouring galaxies with their respective reversal zones. The volumes within these zones have been called the boundaries of the galaxies. The total mass within each domain acts as though it were at the centre of gravity of the domain. Let the total masses within two neighbouring domains be, respectively, m_1 and m_2, where m_2 is the larger.

According to the hypothesis of continuous origin the gross income of a domain is directly proportional to its volume. The volume, in turn, is a direct function of the mass. So the larger, and more massive, domain has the greater income of newly originating matter. If the average rate of origins within both domains exceeds the average rate of extinctions the larger domain will increase in mass and volume more rapidly than the smaller one. The ratio m_1 / m_2, will become a diminishing fraction. If the respective volumes are V_1 and V_2 and the respective distances from the centres of gravity to the reversal zone are D_1 and D_2, V_2 will gain on V_1 at the same time as m_2 gains on m_1, and the greater the discrepancy between V_2 and V_1 becomes the more readily will m_2 gain relatively to m_1. The ratios m_1 / m_2, and V_1 / V_2 will tend at an exponential rate towards zero. The same holds for the ratio D_1 / D_2 In other words the boundary between any two galaxies in domains that contain unequal masses is continuously being pushed towards the centre of the smaller one so long as origins exceed extinctions on both.

This must happen whatever the difference in their masses is, provided only that there be a difference. Even if this is very slight to begin with, the time must inevitably come when the boundary will reach the edge of the smaller galaxy. It will not, however, stop there. It will encroach on the substance of the smaller galaxy and leave this behind it and within the domain of the larger one. Parts of the smaller galaxy will then begin to fall away from it. The process will not even stop when the centre has been reached. It will continue until its whole substance has become forfeit to its more massive neighbour.

For the model based on Asymmetrical Impermanence the rate of origins always exceeds the rate of extinctions, for the latter rate is zero. Hence for this model a domain that is only slightly more massive than its neighbours must always compete successfully with them for hydrogen. It must grow at their expense and thereby become even more massive. In doing so it must also become ever more capable of competing successfully. In the course of time it must swallow every other galaxy in turn. The model based on Asymmetrical Impermanence is thus one that finally consists of but one single concentration of matter. If continuous origin has been going on for all time this model will have an infinite mass and exercise an infinite gravitational pull on new matter. It is not a model that conforms to actuality.

B.2: *Wrong Use of the Concept of Infinity*
Suppose that one could get round the objection to Asymmetrical Impermanence that has just been mentioned. It would not be difficult, for one can always get round an objection to a favoured theory with the help of suitable *ad hoc* hypotheses.

Here the choice might fall on the hypothesis that the inverse square law for force receives a correction such as to limit the distance at which gravitation has a finite value. Galaxies that were further from the biggest galaxy than the limiting distance would then be beyond its range and would not be swallowed up. Thus one would be able to explain the observed existence of discrete galaxies. But even so, this way out of the dilemma could not be allowed, as the following consideration shows.

According to Asymmetrical Impermanence, even when coupled with the further hypothesis that gravitational forces act only over finite distances, the size of every galaxy would be a direct function of its age. Young ones would be small and old ones would grow continuously without limit. Those that had begun an infinite time ago would have become infinitely massive. The potential gradient around them would be infinitely steep. Particles of surrounding hydrogen would be falling on to them with an infinite acceleration. Any atoms of new hydrogen that originated within a finite distance from such galaxies would instantly acquire the velocity of light and become infinitely massive.

The above conclusions are alone surely reason enough for rejecting the hypothesis of Asymmetrical Impermanence. But the reason needs to be explained, for it does not always seem to be properly appreciated. I have pointed out already that a cosmological model can be accepted only if it resembles actuality, and supporters of Asymmetrical Impermanence have claimed, rightly or wrongly, that a model that implies nebulae of infinite mass does not necessarily fail to resemble actuality. All that we can know about the actual universe is, they point out, what is within the optical horizon and therefore observable. The probability that one of the infinitely massive nebulae would occur within this horizon is, it has been said, extremely small. That none are observed does not, therefore, prove that none exist. This was, for instance, the point made by one of the supporters of Asymmetrical Impermanence, F. Hoyle, in a correspondence with myself conducted in the pages of *Nature*. XIV

The argument is of doubtful validity. It holds only if one can postulate that the infinitely massive nebulae are all located at some distance from us such that their infinitely powerful fields at an infinite distance are undetectable here and now. Presumably, then, their distance is some higher power of infinity! However, this is not the only or even basic objection. One must reject a model that includes infinite physical quantities, not because it can be proved wrong by observation, but because it is conceptually wrong. Those who accept this model with equanimity misunderstand the meaning of the concept infinity.

When the symbol for infinity occurs in a mathematical expression it can be translated into words as 'an *indefinite and unspecified* number larger than would define any physical reality with which this expression is concerned'. The words in italics are essential. To say that there can be any time and place where there is a physical quantity for which this symbol represents the correct measure is to misread the symbol.

The meaning of other mathematical symbols is that, when they are replaced by numbers, these numbers will exactly define the quantities that the symbols represent. The numbers will be neither too large nor too small. To imply that the symbol for infinity is similar and can be replaced by a number that is neither too large nor too small is to assume that the symbol stands for a very large but nameable number. Laymen make this mistake every day, but scientists have to learn to avoid it.

B.3: *Growth of Domains for the Model Based on Symmetrical Impermanence*
Thus the model based on Asymmetrical Impermanence must be rejected on conceptual, as well as on observational, grounds. Does the model based on Symmetrical Impermanence fare any better?

I have to admit that I am not sure. If it could be proved that this model is not self-adjusting in the sense of automatically limiting the size of galaxies to finite values I should reject it at the cost of sacrificing the great explanatory

---

XIV *Nature,* Vol. 165, pp. 68 and 687, on respectively 14 January and 29 April, 1950.

power that has been demonstrated for this hypothesis, both in the main parts of this book and in the appendices that are to follow. Only a searching mathematical study can show whether it is so and this study is among the many that I prefer to leave to others. I shall rest content to show here how a superficial investigation is encouraging and suggests that Symmetrical Impermanence is likely to stand this test, as it has stood all the others to which I have subjected it.

If the size of a galaxy is to be self-adjusting extinctions within its domain must equal, or exceed, origins when the galaxy has reached a certain size. This can only happen when the average mass density in the whole domain equals or exceeds the equilibrium value. The quantity dm / dt for the domain will then be negative.

This would happen if the reversal zone around an old and massive galaxy were, from time to time, to be pushed inwards by the new galaxies that grow around it. Such a process would transfer volume from the domain of the old and more massive galaxies to the domains of the newer and less massive ones. As much of the mass within a domain is concentrated in the galaxy at its centre the loss of volume of the older domain would not be accompanied by a proportional loss of mass, so the mass density would increase. If it did so until the equilibrium density was exceeded the galaxy would dwindle.

Let the average mass densities in two neighbouring domains be, respectively, σ_1 and σ_2. If the model is to be self-adjusting the ratio σ_1 / σ_2 must be a direct function of m_1 / m_2 or V_1 / V_2. For then the more massive and more voluminous domain would have the greater mass density. Its loss by extinction would be relatively the greater. If it is so the reversal zone will move at times away from the incipient and towards the older galaxy until extinctions have brought the density to below the equilibrium value.

One can only know if this happens after one has calculated the way in which the gravitational fields of a large assembly of galaxies are superimposed. In other words the contours of the astronomical landscape and the changes that it undergoes must be known with some precision. What is known at present is too vague to permit the question to be answered with any certainty. But one may approach an answer by a process of successive approximations. Suppose one begins by ignoring the fields of all galaxies except two, an incipient one and one of its older neighbours. One will then define the position of the reversal zone between them by the approximate equations given in Chapter 10.

$$(D_1 / D_2)^2 = m_1 / m_2 \ldots \ldots (10a)$$

If one puts as a further approximation,

$$V_1 / V_2 = (D_1 / D_2)^3$$

one can write

$$V_1 / V_2 = (m_1 / m_2)^{3/2}$$

from which

$$\sigma_1 / \sigma_2 = (m_2 / m_1)^{1/2} \ldots\ldots\ldots (Ba)$$

If this were accurate the larger mass would be in a domain with the smaller mass density. The rate of extinctions per unit volume would be smaller in the domains with the smaller volume. The model would be the reverse of self-stabilizing. But equation (lOa) cannot be very near the truth.

One approximates more closely to the truth when one takes a third galaxy into account. This has been done in Chapter 12 with equations (12c) and (12f). But these two are not quite correct. A further approximation has been adopted there by substituting an astronomical pass for a summit.

The potential gradient near the pass is approximately

$$E_2 = -4Gm_2r / D^3$$

derived from (12f)
where D is the distance from the reversal zone to the older galaxy with mass m_2 and r is the distance from the reversal zone to the astronomical summit before a new cloud has begun to form there. Let this have formed and have acquired mass m_1. The potential gradient attributable to m_1 only is then

$$E = Gm_1 / r^2$$

At the reversal zone the two gradients sum to zero and one can write

$$Gm_1 / r^2 = 4Gm_2r / D^3$$

from which

$$m_1 / m_2 = 4(r / D)^3$$

The ratio of densities is then

$$\sigma_1 / \sigma_2 = 4$$

The further approximation gives four times the density of the older domain to that of the incipient cloud so long as r is small. If this were the true relation the density in the domain of the older galaxy would be one quarter of the equilibrium value when this value was reached by the incipient cloud. The older galaxy would continue to grow more massive and would extinguish the incipient one. The objection already found against Asymmetrical Impermanence would also be valid against a Symmetrical Impermanence.

However, equation (12f) is still misleading, if not as much so as equation (lOa). As has been pointed out at the end of Chapter 12, an equation that gives the potential gradient around an astronomical summit, as distinct from one at an astronomical pass should take some such form as

$$E = - \{ GmDr / (D^2 - r^2)^2 \} \varphi(r / D) \ldots\ldots \ldots\ldots \ (12g)$$

By this nearer approximation the position of the reversal zone may perhaps be defined by

$$Gm_1 / r^2 = -\{Gm_2Dr / (D^2 - r^2)\} \varphi (r / D)$$

from which

$$m_1 / m_2 = - \{ D r/(D^2 - r^2)^2 \varphi (r / D)$$

If r is very small compared with D this gives

$$\sigma_1 / \sigma_2 = -\varphi(r / D) \ldots\ldots\ldots\ldots \ (Bb)$$

As an astronomical summit is flatter than a pass the function varies directly with r/D, as was said in Chapter 12. Hence it is more than probable that the function is such as to cause the smaller domain also to have the lower mass density. If so the possibility need not be excluded out of hand that the mass density in the larger and older domain may exceed the equilibrium value while that in the domain of the new galaxy is below it.

It is, however, not sufficient for $\varphi(r/D)$ to be the right kind of function in order that the size of galaxies may always be finite. It is also necessary for the half-life of matter to be below a certain value. If the half-life of matter were infinite and matter were originating continuously as advocates of Asymmetrical Impermanence assert the whole of the material universe would form a single infinitely massive concentration whatever form was taken by $\varphi(r/D)$. What the half-life must be depends on the nature of $\varphi(r / D)$, but some general considerations will be given in Appendix C, which show that, by astronomical standards, the half-life must be less than one might expect. Further considerations to be given in later appendices will, however, show that a rather short half-life is consistent with sundry well-known facts of observation. Astrophysicists, geologists and biologists will indeed have reason to welcome a rather short half-life. For it helps to explain several facts that have hitherto defied explanation.

The conclusion that is hinted at by equations (Ba) and (Bb) can be expressed differently as follows. If there were only two galaxies the larger one would inevitably prevail by competing successfully with the smaller one. But with a three-dimensional arrangement the time comes when an old and large galaxy is surrounded on all sides by small new ones. It is suggested that their combined competitive power suffices for their domains to encroach successfully on the older one. If the half-life of matter is short enough the encroachment suffices to increase the mass density of the large galaxy to above the equilibrium value, leading thereby to a reduction in its mass.

B.4: *The History of a Domain*
Let us consider, in view of the automatic adjustment that seems to follow from equation (Bb), how the mass of a domain must vary with time. There will be a moment when neighbouring domains contain equal masses. Their average mass densities and volumes must then also be equal and the astronomical pass between them must be equidistant from their respective centres.

In expanding space the centres drift apart and the domains grow more and more voluminous. Therewith they receive new hydrogen at an increasing rate and become more and more massive. The new hydrogen falls on to the galaxies at the centre of the domains and is added to their masses. But only when the hydrogen completes the journey. Some of it must, according to continuous extinction, become extinct while it is under way. The proportion that does so must increase with an increasing average journey, and so an increase in volume of a domain does not result in a proportional rate of increase of the mass of the galaxy at its centre. But, nevertheless, the rate at which this galaxy grows by capture does vary directly with the volume of the domain.

This volume does not grow indefinitely. The time must come when the potential gradient around the neighbouring astronomical summits falls to the value at which a new cloud can begin to form. When this happens the new cloud competes with the surrounding galaxies for hydrogen. In other words its domain encroaches increasingly on theirs. The older domains then begin to dwindle in volume and mass. But the regions that they lose are the outermost ones and these are very tenuous; the dense central core remains in the old domain. Hence the average mass density of the old domain is greater than corresponds to its size and begins to be reduced by a preponderance of extinctions over origins. The adjustment takes time and so there is a time lag, a phase displacement, between the process of reducing the volume and of reducing the mass. But this need not concern us at the moment.

210

Eventually the new cloud becomes a new galaxy, as massive as its neighbours. It and they drift apart. Their domains grow and therewith their incomes. They begin to grow again.

It is not difficult to understand that a new cloud must begin to form on an astronomical summit as soon as the gradient has fallen to the critical value, and that this must happen in time for every astronomical summit in extragalactic space. For the gradient continues to jail while space expands and the neighbouring galaxies drift further away.

Clearly the average distance between two neighbouring galaxies at the moment when a cloud begins to form must have a constant value D_{av} and the average distance of the incipient cloud from the two galaxies is half this value or $D_{av}/2$. When the cloud has itself become a galaxy and its distance from the neighbours has grown from $D_{av}/2$ to D_{av} a new cloud will begin to form on the new astronomical summit that has developed.

Thus the time interval between successive generations of galaxies must average the time taken for the linear dimensions of space to double themselves. This has been given as three-and-a-half thousand million years. But it depends on the value of Hubble's constant, about which there is some uncertainty at the time of writing. This time is the period of fluctuation in the mass of a galaxy.

Appendix C

The Half-life of Matter

C.1 : *The Need for Quantitative Thinking*

The reasoning presented in this book has purposely been predominantly qualitative and this because, as was explained in Chapter 12, one cannot usefully do much quantitative work on any problem until one has reached clarity about the concepts concerned. A thorough understanding of the meaning of the algebraic symbols that are to be used must precede mathematical treatment if this is to be profitable. If mathematics is applied prematurely, it becomes no more than an idle exercise; it remains unrevealing; it may mislead.

But, nevertheless, quantitative work should, where possible, always follow the search for meaning. That quantitative thinking should take second place in order of procedure does not imply that it necessarily takes second place in order of importance. For this reason a small amount of mathematics has proved to be unavoidable. Its result has been reassuring. But much more will have to be done in order to establish the Principle of Minimum Assumption as a sound basis.

Further tests of this principle, be they quantitative or qualitative, will involve many disciplines other than astronomy; any branch of physics may be able to contribute something; a great variety of different mathematical methods may have to be employed; a range of knowledge, skill and experience will be required that cannot be possessed by any one scientist. So I have to rest content to leave the task of further testing and development to others. But before doing this I should like to draw particular attention to the importance of the cosmic constant that I have called the half-life of matter.

According to the traditional hypothesis (Bl), at the beginning of Chapter 3, an elementary component of the material universe never becomes extinct. At most it is converted from one form into another. This is equivalent to saying that the half-life of matter is infinite. But according to the Hypothesis of Symmetrical Impermanence the half-life of matter is finite. It was mainly quantitative considerations that caused me to reduce the half-life from infinity to an unspecified finite value. Before Symmetrical Impermanence can become established, further quantitative considerations must bring this value between definable lower and upper limits.

It appears that such limits can be arrived at by several distinct lines of reasoning, each starting from different known facts. If the Hypothesis of Symmetrical Impermanence is valid, all these limits must leave a bracket of possible values. The lower limit set by one approach must not be higher than the upper limit set by another. Every effort must be made, moreover, to narrow the gap between the lowest and the highest of the possible values. Let me give a hint here as to ways in which these limits may be estimated.

C.2: *List of Symbols*

$N_a =$ Average number of elementary components per unit volume in a fair sample of the universe. (A fair sample may have to be extensive and to include many extragalactic nebulae; for the mass density in the domain of any particular nebula may differ significantly from the average value.)

$N_{eq} =$ Number of elementary components per unit volume at the equilibrium density. (This is the density at which the rates of origins and extinctions per unit volume are equal.)

$N_o =$ Number of elementary components that originate in unit volume and time.

$n_e =$ Fraction of elementary components in any volume that become extinct in unit time.

$T_s =$ Time during which the linear dimensions of space double.

$T_m =$ Half-life of matter.

$p =$ T_m / T_s

$q =$ N_{eq} / N_a

$m =$ Mass of the earth at any time t

$m_o =$ Mass of the earth when $t = 0$

C.3: *The Earth's Loss of Mass*
According to Symmetrical Impermanence every body is losing mass at a rate determined by the half-life of matter. But bodies like the earth, the moon and the sun are also gaining mass by capturing it from their surroundings. However, the sun must compete successfully with the earth and the moon for any matter that is in process of falling towards the solar system. Such matter must nearly always fall past the planets and their satellites and on to the sun.

Hence it would need some research before one could say whether, on balance, the sun is gaining or losing mass, whether dm / dt is positive or negative. But one may feel sure that the earth is gaining only a negligible quantity and that, according to Symmetrical Impermanence, dm / dt must be negative and determined to a very close approximation by the half-life of matter.

This can be expressed by the equation

$$dm / dt = -n_e m$$

from which

$$m = m_o e - {}^n_e{}^t$$

so

$$m_o = m e^n_e{}^t$$

and

$$\ln(m_o / m) = n_e t$$

For the half-life of matter

$$t = T_m \text{ and } (m_o / m) = 2$$

Hence

$$\ln 2 = 0.69 = n_e T_m \text{ .... (Ca)}$$

It follows that for any values of m_o and t

$$\ln(m_o / m) = 0.69 t / T_m \text{.......... (Cb)}$$

For small values of t/T_m this can be replaced by

$$(m_o / m) = 1 + 0.69 \, t / T_m \text{ .... (Cc)}$$

From this equation it is seen that the loss of mass of the earth during the last 1,500 years has been one part in ten million if T_m is 10^{10}, one part in a million if T_m is 10^9 and one part in a hundred thousand if T_m is 10^8.

The value of g has diminished correspondingly. Any calculations that have assumed g to have remained constant during 1,500 years must therefore be in error by a corresponding small fraction.

Eclipses are among events affected by such calculations. If the earth was more massive in the past than it is now, it had also a firmer hold on the moon. This

214

body was nearer to the earth and completed its orbit in a shorter time than it would have done if g had been no greater than it is today. If the records of past eclipses were completely accurate, and if all other quantities that enter into the calculation of an eclipse were known with absolute precision, one could compare recorded times and places with calculated ones. If the calculation assumed that g had remained constant, any difference between records and the result of calculations would then reveal the change that g had undergone.

But records cannot have been so very precise 1,500 years ago and calculations made today have to allow for a small and unavoidable margin of error. So a small change in the value of g would be masked by various uncertainties. Among the circumstances that significantly affect the time and place of an eclipse tides are important. They cause the earth to rotate more slowly while loss of mass by extinction causes it to rotate faster. For this reason a rather large change in the value of g would be necessary for a comparison of recorded and calculated eclipses to reveal it.

Only experts in astronomical measurement could tell us how great a rate of change of g could be detected with the help of all available data. If no such change has in fact been detected, one may safely conclude that the rate of change is less than this value, though one cannot know how much less. The negative evidence that no change has been detected would thus set a lower limit to the half-life of matter. It seems probable that a detectable rate would have to be much greater than corresponds to a loss of mass of one part in a million during 1,500 years, and so the lower limit set by this consideration is probably much less than 10^9 years.

A small change during 1,500 years is, however, equivalent to a large one during the whole of the earth's existence and it may be possible to find reasons why the earth cannot have been more massive at its beginning than some definable value. I do not know of any limiting consideration and yet I do not like to contemplate the notion that the earth has ever been many times as massive as it is today. Probably my disinclination to do so is emotional and therefore inadmissible in a scientific inquiry. So let any objection to a large value for the initial mass of the earth be ignored unless it can be more rationally justified.

At the time of writing this the estimate for the age of the earth that is given by the radium clock and seems to have the widest support is 2.6×10^9 years. Reasons for reducing this estimate greatly will appear later, but if one gives this value to t in equation (Cb) one obtains the following table. In this m_o / m is the ratio of the earth's initial to its present mass and T_m is the half-life of matter:

Table 1

$T_m \times 10^9$ years	m_o / m
0.3	395
0.4	89
0.5	36
0.6	19
0.7	12
0.8	9
0.9	7
1.0	6

It is doubtful whether anyone knows anything at all about the mass of the earth, its condition or its other properties at the time when it began. So any figures concerning its mass at that time can neither be proved true or false.

Knowledge begins at a date that may roughly be put at two thousand million years ago and so it is relevant to consider what the maximum possible value of the mass of the earth can have been at that time. If the half-life of matter is 10^9 years, its mass two thousand million years ago was about four times its present mass. Its radius was therefore about 60 per cent greater than it is now. But reasons will be found in this appendix, as well as in Appendices D and E, why this is probably a gross under-estimate. Any emotional distaste one may experience for the notion that the earth has today but a small fraction of its initial mass must yield to some rather cogent facts, which are partly astronomical, partly geological and partly biological.

C.4: *The Rate of Extinctions per Second*
According to the new theory of gravitation that has been presented here, the gravitational field is quantized. The gravitational force occurs in jerks. But these are not observed; the force seems to be continuous, not only for the attraction exercised by the whole earth but also for that exercised by those leaden spheres that have been used in the measurements of the gravitational constant G by Cavendish, Boys and Poynting. It is evident that the jerks follow each other in quick succession and as each jerk is the consequence of an extinction the rate of extinctions in a sizeable lump of lead must be high. This consideration sets an upper limit to the value of T_m.

From equation (Ca) the fraction of a given mass that becomes extinct in unit time is $0.69T_m$. From this it is easy to find that the fraction that becomes extinct per second is $2.2 \times 10^{-8} / T_m$ if T_m is measured in years. A gramme has the mass of about 6×10^{23} protons and if the elementary component that becomes extinct has the mass of a proton the number of components becoming extinct in a gramme during every second is $1.3 \times 10^{16} / T_m$. If T_m is 10^9 the number of extinctions in one second is over ten million, much more than sufficient to make the field appear to be continuous. It follows that a mass of one ton of hydrogen would produce quanta of gravitation at a rate of over one per second if the half-life of matter were 10^{24} years. It is evident that a very long half-life is consistent with the apparent continuity of the gravitational force.

C.5: *The Equilibrium Density*
During the time T_s that it takes for the linear dimensions of space to double the volume of space increases eightfold. If the average density N_a is to remain constant there will be seven new elementary components at the end of time T_s, for every one that there was at the beginning of that time.

During the time T_s, the number of elementary components becoming extinct is $n_e N_a T_s$, and so the gross number of origins is

$$N_o T_s = 7N_a + n_e N_a T_s \text{ .... (Cd)}$$

From this it follows that in a fair sample of the universe the ratio of origins to extinctions during unit time is

$$(N_o / n_e N_a = (7 / n_e T_s) + 1$$

Substituting for n_e on the right hand side from equation (Ca) one obtains

$$(N_o / n_e N_a) = (7T_m / 0.69T_s) + 1$$

Putting $T_m / T_s = p$ *then gives*

$$(N_o / n_e N_a) = 10p + 1 \text{ nearly.}$$

Now at the equilibrium density, N_{eq} the rates of origins and extinctions are equal giving $(N_o / n_e N_{eq}) = 1$, from which it follows that

$$(N_{eq} / N_a) = q = 10p + 1 \text{ .... (Ce)}$$

This expression enables us to place an upper limit to T_m, It has been shown in Appendix B that a galaxy would grow without limit if the total number of origins within its domain always exceeded the rate of extinctions there. For the size to stabilize there must be periods of time during which the rates are reversed and during these periods the average density in the domain must exceed the equilibrium value N_{eq}.

What causes the density to rise above N_{eq} is that the domains of the new clouds that are growing around the older galaxy spread and encroach on the

217

domain of the older galaxy. Thereby the latter loses those regions in which the density is lowest, while it retains the central region in which the density is high. While the mass within the domain remains almost unaltered, the volume of the domain is reduced.

If the boundary of the domain of the new clouds were to advance from all sides up to a mid position between the centres of the new clouds and the older galaxy, the domain of this would be reduced to about one-eighth of its original volume. But the new clouds are much less massive and so the encroachment cannot go so far.

The reduction of the volume of the older galaxy cannot, of course, be sufficient to leave only one-eighth of the previous volume. At the same time the reduction must be great enough to raise the average mass density from less than N_a to more than N_{eq} , for if it were not so, if the density in the reduced volume still remained below that given by N_{eq}, the rate of origins within the domain would always exceed the rate of extinctions and the older galaxy would continue without interruption to become ever more massive.

We thus have to conclude, firstly, that fluctuations in the domain of any galaxy range over less than one-eighth in volume, and secondly, that the ratio N_{eq} / N_a = q is less than the ratio of the maximum to the minimum volume of the domain.

It would be a complicated matter to calculate the ratio of maximum to minimum volume that should be expected from the growth of new clouds around an existing galaxy, though this will have to be done some day. Meanwhile, Table II will suffice to show how q and T_m are interconnected by equation (Ce). In preparing the table it has been assumed that T_s is 3.66 x 10_9 years. As p is the ratio of T_m to T_s the half-life of matter T_m is obtained from equation (Cc).

Table II

q	T_m years
5	1.46 x 10^9
4	1.10 x 10^9
3	0.73 x 10^9
2	0.37 x 10^9

Table I makes it appear unlikely that T_m could be much less than 3 x 10^8 years. It is unlikely that q can be much greater than 2 and so Table II makes it appear unlikely that T_m can be much greater than 4 x 10^8 years. If various lines of approach do not allow much room for manoeuvre, they do not at least lead to contradictory conclusions.

It will be shown in Appendix E that a different approach also suggests a value of the order of 4×10^8 years. But if the age of radio-active substances is found to be much greater than 2.6×10^9 years there may be a reason for assigning a longer half-life to matter and, on the other hand, evolutionists may be influenced by the conclusions reached in Appendix E to press for a shorter estimate.

The conclusion that the half-life of matter is shorter than that of some radio-active substances may cause surprise, perhaps be even thought to be impossible. The reason why this conclusion need not be inconsistent with observed facts can only be given after the relation between space and mass has been explored and will be found at the end of Appendix H.

Appendix D

Origin of the Planetary System

D.l: *A Star's Gain and Loss of Mass*

Continuous origin is a diffuse process. As its rate per unit volume is everywhere constant, this rate is also low; hence clearly observable effects cannot be expected over short distances. To be pronounced they must occur in extragalactic space. This explains why the best evidence in support of continuous origin is provided by the spiral nebulae.

Continuous extinction, on the other hand, has its highest rate per unit volume where matter is most concentrated, namely in the heavenly bodies. Its observable effects are localized. This is why the best evidence in support of it is provided by gravitation.

But gravitation cannot be the only manifestation of continuous extinction. One of its implications is that the earth and the other planets, together with their satellites, are continuously losing mass by extinction. If T_m is the half-life of matter, the masses of these bodies must have been twice their present values T_m years ago and will have half their present values T_m years hence. Such a change must have observable consequences.

It must be noted that the sun's mass cannot have experienced a change at the same rate. For while the sun is assumed to be losing mass by extinction, it is experiencing two additional changes. It is losing further mass by radiation and it is gaining mass by the capture of hydrogen from its surroundings. The net result may be either a loss or a gain. It would not be easy to estimate which in fact it is.

What the planets gain by capture must be negligible. Its great mass makes the sun a successful competitor for new matter. The sun's domain is large compared with that of any planet; in other words a particle that finds itself at only a comparatively short distance from a planet does not fall on to the planet but on to the sun. Hence the sun may have been either larger or smaller in former times than it is now, but the planets must have been larger. If their loss of mass were known, it could be used to evaluate T_m.

If any observed facts showed that there has been no loss of mass, one would have to conclude that T_m was infinite; which would be to abandon the hypothesis of continuous extinction. But I do not know of any such facts. On the contrary, wherever I turn I find facts that seem to give a finite value to T_m They are all consistent with a value less, perhaps substantially less than 10^9 years.

These facts belong to several disciplines, including astronomy and biology. They are facts that have hitherto defied a satisfactory explanation. But then those who tried to explain them took it for granted that the half-life of matter was infinite. I propose to show in this appendix and in Appendix E that there is a good prospect of finding explanations of many baffling facts if one adopts the hypothesis of continuous extinction.

A detailed attempt at solving riddles that have been presented by facts in these various disciplines would be beyond the scope of the present study. It will have to be left to the combined efforts of various experts. I shall be content here to give no more than a bare outline of some further conclusions that are reached on the basis of Symmetrical Impermanence. I am not claiming to give the result of an exhaustive study but, instead, some tentative conclusions. I am hoping that they are at least near to the truth and that they will suggest rewarding fields of research for some astro-physicists and biologists. What is presented in these Appendices D and E might aptly be regarded as outlines of some research programmes.

D.2: *The Double Star Theory of Planetary Formation*
In the past many theories have been invented in order to explain the origin of the planetary system. Each in turn has proved defective in some particular, has been abandoned and replaced by another one. One of these theories was developed a few years ago in sundry detail by F. Hoyle. It is based on a suggestion that seems first to have been made by Lyttleton. Hoyle has put forward a quite different theory since then; so he presumably has found the earlier one to be untenable, as indeed it is. But its core had so much to recommend it that the earlier theory ought not, I think, to be wholly abandoned. There is a case for retaining the core and correcting the errors. I propose to show here that symmetrical impermanence provides a means of doing this. Hoyle's earlier theory was as follows.

At one time the sun had a companion with which it formed one of the numerous double stars that are known to astronomy. Both stars consisted mainly of hydrogen, as most stars are known to. In both the pressure and temperature were such that the hydrogen was being slowly converted into helium, as it still is in the sun. The large quantity of energy released by this conversion deep in the interior of the stars appeared as radiation and exerted sufficient pressure on the gas particles above them to push them outwards against gravity. The internal radiation acted, in other words, like air in a balloon and kept both stars inflated. It is generally accepted today that this does happen in all stars in which the rate of helium synthesis is between certain upper and lower limits, and is still happening in the sun. The diameter of the sun would be a small fraction of its present value if the helium factory did not maintain a considerable internal pressure of radiation.

According to Hoyle's double star theory the time came when a large proportion of the hydrogen in the sun's companion began to be exhausted, while the sun still had enough of the factory's raw material, hydrogen, to maintain internal pressure of radiation and has enough even now for a long while to come.

Failing to maintain its previous internal radiation the sun's companion collapsed under the influence of its own weight. Thereby the nuclei of the remaining hydrogen and of the helium that was present were forced into very close association. The pressure and temperature were sufficient to cause the synthesis of all the heavier elements. The energy required for this synthesis was largely derived from the potential energy released in the star's collapse. After the synthesis the star consisted to a smaller extent of hydrogen and helium. Most of its substance was the same as is now found in the planets.

Hoyle proceeded from there to some further assumptions. He said rightly that in collapsing the angular velocity of the sun's companion increased. By the principle of conservation of angular momentum it would have to do so. But Hoyle went on to assert that the accompanying increase in centrifugal force sufficed to exceed the force of gravity, that wisps of gas were thrown off from the star, that these subsequently condensed into the planets, that centrifugal force thereafter propelled the parent star out of the sun's neighbourhood. It is these further assumptions that need to be corrected.

The core of the theory carries conviction and would make it far superior to any rival if it were consistent with known facts in other respects. Other theories have assumed that the planets were formed out of the sun's substance. But of the elements up to uranium, with a nucleus that carries 92 charges, hydrogen and helium are the only ones that occur with any abundance in the sun. Of the others there are, at most, only traces. If the planets had been formed out of the sun's substance, one would have to find a reason why the substances are so different. One might succeed in inventing an hypothesis by which the synthesis of the heavier elements had been formed after they left the sun. But no satisfactory reason why this should happen has ever been found, and the hope of finding one seems remote. The conditions needed for converting lighter into heavier elements are now well understood and they require a combination of very high pressure and temperature. It is impossible to believe that they can have occurred in the comparatively small bodies that formed the planets.

Alternately, one might perhaps postulate some filtering process by which the heavier elements would be extracted from the sun and the lighter ones left behind. Ingenuity can accomplish much. But it must be allied with disciplined thought if it is to achieve anything of lasting value. The notion that all the elementary substances existed in the parent body from which the planets were formed, and in roughly the same relative quantities, is less fantastic than any other that seems to have been put forward. It deserves to be tested for its consistency with other known facts. Many doubts and queries arise. Let each of them be considered in turn.

A reason has to be found why the two companion stars should have had such different fates. What circumstances could have left the sun with enough hydrogen to go on synthesizing helium for many thousands of millions of years while its companion's store was exhausted at an early date? One possible answer would be that the sun's companion used up its hydrogen more quickly than the sun did, another is that it had less hydrogen to use up.

Advocates of the earlier double star theory have chosen the former explanation. Their view is that for some reason the helium synthesis was greatly increased at a certain moment in the star's life. It is reasonable to suppose that this happened, for the synthesis can become an unstable process and would, if it increased above a certain critical rate, get out of hand. The subsequent collapse is assumed to have been correspondingly catastrophic and to have occurred in a very short period of time. The potential energy of position possessed by the substance of the star in its inflated condition was converted into heat at the surface, where it would have caused a very great increase of brightness.

It is believed by some that a supernova is the result of this kind of catastrophic collapse of a star, and so advocates of the double star theory believe that the substance of the sun's companion was converted into the heavier elements during the process that is manifest as a supernova.

The second explanation, namely that there was less hydrogen to begin with, provides a less dramatic history but is more consistent with Symmetrical Impermanence. It would seem equally well to account for production of the heavier elements.

If the sun's companion were somewhat less massive than the sun, it will have lost to the sun in competition for hydrogen from the surroundings. While the sun could replenish its store as fast as it was being consumed, the smaller companion could not do so. Eventually the moment must then have arrived when there was not enough fuel to maintain the amount of radiation needed to inflate the star. Consequently it collapsed. According to this explanation, the pressure after collapse would be the same as for the history in which the hydrogen was being used up in a burst of extravagance, and the temperature also could hardly be very different. One should expect, with the assumption of gradual consumption, the combination of pressure and temperature to account equally satisfactorily for the synthesis of the heavier elements. Tentatively I should therefore like to put forward the theory of hydrogen starvation in preference to the current one of an orgy of spending as an explanation of the exhaustion of the hydrogen store. But a decision can only be reached by a specialist in this branch of astrophysics. Whether the hydrogen consumption and collapse were slow or rapid, the collapse must have been accompanied by a great increase in the rotational speed of the star. For by the principle of conservation of angular momentum the angular speed increases as the moment of gyration decreases. During collapse, therefore, the centrifugal force on the substance of the star must have increased to a very high value.

D.3: *Disruption by Centrifugal Force*
The effect of a centrifugal force that reached such a high value needs to be thought out with care. It is fatally easy to jump to the conclusion that the star would necessarily break into fragments like a bursting flywheel and fling the pieces away to a great distance. But the truth is that this would not happen to a gaseous star. From what I have read and heard said I gather, nevertheless, that there is some misapprehension about this, so I make no apology for presenting here some elementary mechanics.

A gaseous star consists of a collection of separate particles. In the outer layers these are molecules of the gas and there is no cohesion between them. They have momentum and are subjected to the forces of impact and gravitation. Centrifugal force acts on them in the same way as it would act on loose stones resting on a smooth surface. As a step towards under- standing the effect of centrifugal force on the gas particles, let us therefore first consider what the effect would be if the star were solid and an isolated stone were lying on its horizontal surface.

The star is supposed to be rotating faster and faster. There is thus a tangential acceleration of its surface and the stone, having inert mass, would be left behind if the surface on which it rests were perfectly smooth. But this need not worry us for the moment. Let the surface be rough enough to entrain the stone by friction. It will then share the acceleration of the surface of the star.

As the angular velocity increases, the apparent weight of the stone decreases. The moment arrives eventually when the pressure between the stone and the surface on which it rests becomes zero. If the angular velocity of the star does not increase further beyond this value, the stone will continue to retain its position on the ground by virtue of its inertia. If the speed of rotation increases still further, the stone will, however, not be further entrained by friction; for there is no friction. It will continue to move with its former velocity, which is now less than that of the surface on which it has been resting. To an observer on the star it will appear to slip over the surface in the opposite direction to that of rotation. But so long as it is free from restraint nothing will cause the stone to rise from the surface.

Now let it be supposed that there is restraint. Let the stone be held in position by a wire anchored to a bolt in a rock. By this means the stone is prevented from slipping backwards as the star accelerates. It is made to participate in the rotation of the star, whatever this may be, for so long as the wire holds. When the speed of rotation exceeds the value at which gravity is exactly balanced, the stone will, by virtue of its inertia, tend to move with constant velocity in the straight line that forms a tangent to the surface of the star instead of following the curved path taken by the rock to which the wire is attached. As the distance between the tangent and the circumference increases, the stone will be seen to rise. It will do so until the wire prevents it from rising further. The wire will then be stretched vertically upwards with the stone at its end. It will look like a captive balloon.

If the speed of rotation increases further without limit, the centrifugal force on the stone will eventually exceed the tensile strength of the wire. The stone will then continue to move tangentially by virtue of its inertia. As the distance between tangent and circumference increases, the stone will seem to be rising vertically. But it will now be free from centrifugal force and again subject to gravity. Superimposed on the tangential move- ment at constant velocity there will be a radial movement downwards with the acceleration of gravity. It is not difficult to show that the vector sum of these movements, if the force exceeds a certain critical value, is an orbit around the star.

The diameter of the orbit depends on the excess of centrifugal force over gravity at the moment when the wire breaks and this, in turn, depends on the tensile strength of the wire. As has already been shown, the stone does not leave the ground if there is no wire.

Gases and liquids have no tensile strength. Particles of either are like stones not secured by a wire. Centrifugal force could never throw off puffs of gas or drops of water from a rotating star, however fast it rotated. For any substance to leave the surface there must be cohesion, the equivalent of a securing wire. The tensile strength of a solid substance has to be reached so that the material that breaks off may have sufficient momentum to leave the star and pursue an independent orbit.

This conclusion is not inconsistent with the double star theory but amplifies it. If the planets came out of a companion of the sun, they can only have done so after the companion had changed from a gaseous to a solid state. The star's substance may have been plastic enough to be deformable, but it must have had cohesion.

D4: *Can Centrifugal Force ever cause a Star to throw off Fragments?*
It does not, however, save the double star theory to do no more than to postulate that the sun's companion became solid before it hurled the planets off into the surrounding space. There are several further serious difficulties, one of which is that any star, solid or gaseous, would not be disrupted by centrifugal force, it would not hurl fragments into space, so long as it retained its original mass. From the views that have been expressed by some I have had to conclude that this is not understood as generally as it needs to be. So I make no apology for discussing here some elementary facts concerning the relation between cooling, contraction, cohesion, rotation, centrifugal force and gravity.

Gravity is a centripetal force and its direction in a star is opposite to that of centrifugal force. A particle anywhere within a rotating star is subjected to both forces and the direction of its movement depends on which force exceeds the other in magnitude. If gravity is the greater and the particle is unrestrained, it will move towards the centre. If centrifugal force is the greater, it will move away from the centre.

Consider a particle of mass m in a star of density σ and rotating with an angular velocity ω. Let the distance of the particle from the centre of the star be r. The mass, m within radius r is $(4/3)\pi\sigma r^3$ and the gravitational force towards the centre of the star is

$$F_g = (4/3)\pi G\sigma Mr.$$

The centrifugal force is

$$F_c = \omega^2 Mr.$$

When these forces balance one has

$$\omega^2 = (4/3)\,\pi G\sigma \,\ldots\ldots\ldots\,\text{(Da)}$$

This shows that the angular velocity at which centrifugal force equals gravity depends only on the density of the star. If this is uniform from the centre outwards, and if all parts of the star have the same angular velocity, a balance in one place means a balance in every place. When the angular momentum defined by equation (Da) has been reached, there is no tendency for any particle, large or small and wherever situated, to move outwards towards the circumference or inwards towards the centre. The motion of particles is determined only by random collisions.

If in such a star there is pressure of radiation, its effect is equivalent to a reduction in the value of G. It does not alter the form of the equation.

When the angular velocity has reached the limiting value defined by equation (Da) it can never increase further so long as gravity alone is the cause of contraction. For the star will cease to contract. No particle can weigh on others so as to press them deeper into the star's interior. There is, indeed, no predominant vertical pressure, either downwards or upwards. There is nothing to hold particles tightly against each other. The star could be cooled down to within a few degrees absolute and its component parts would still fail to rest on each other, as grains of sand do when piled into a heap. The squeezing together of particles can only occur if ω is smaller than defined by equation (Da). And disintegration can occur only if ω is larger.

Bearing this in mind one should not expect a cooling star to solidify if its angular velocity were as high as is defined by equation (Da), for the molecules would then not be squeezed into the condition of interlocking that characterizes the solid state. Hence the double star theory can hardly be saved unless one postulates a low value for the angular momentum of the sun's companion. This must have been low enough for gravitation to predominate over centrifugal force when it was solidifying.

To save the double star theory one must assume, let me repeat, that the angular velocity of the sun's companion was low enough after its collapse for gravity significantly to exceed centrifugal force. If it was not so there would have been no compressive forces within the star; nuclei would not have been brought close enough together for the heavier elements to be synthesized. But if gravity exceeded centrifugal force no fragments of the star would be thrown off. So one has also to assume that, subsequent to solidification, centrifugal force significantly exceeded gravity. For this to have happened the angular velocity of the star must have increased very greatly after the star had been compacted into a solid mass. Could this have happened?

The answer is, of course, that according to traditional views it could not. In the absence of an external couple the angular momentum of the star must be conserved and so an increase in angular velocity must be accompanied by contraction. But traditional views seem to permit only one cause of contraction. This is cooling coupled with gravitational force.

The temperature coefficient of solids is not great and so thermal contraction cannot be great after the star has solidified, even when there is a centripetal pull on the component particles. But it has to be remembered that this pull ceases as soon as centrifugal force balances gravity. Thermal contraction, however great, cannot lead to an excess of centrifugal force over gravity. To save the double star theory a different cause of contraction must be found. The contraction must be so considerable that a strong preponderance of gravity over centrifugal force is replaced by a sufficient preponderance of centrifugal force over gravity to overcome the tensile strength of the material of which the star is made and to cause fragments to be torn off and buried away.

If I were called upon to invent an *ad hoc* hypothesis to explain so great a contraction, I should not attempt the task. I should instead remember Newton's 'hypotheses non fingo'. But an *ad hoc* hypothesis is not necessary. The contraction is an inference from Symmetrical Impermanence.

D.5: *The Effect of Continuous Extinction on Centrifugal Force*
The sun's companion, it is assumed in the revised double star theory, was the smaller partner and gained less and less mass by capture as its relative size decreased more and more and its competitive capacity relative to the sun decreased correspondingly. In the course of time the sun's companion therefore lost mass and volume and continued to do so after its solidification. So long as gravity exceeded centrifugal force the loss was accompanied by contraction without limit. The density remained roughly constant while the radius decreased. Any given particle with mass m moved from a radius r to a radius r_1, less than r. In conserving its angular momentum it acquired an angular velocity ω_1 greater than the previous value ω.

This could continue only until the condition defined by equation (Da) was reached. Thereafter extinctions resulted in a reduction in the value of σ with no corresponding change in the value of r and ω. The consequence would be a decrease in gravity for a particle in any given position. The star would become too flimsy to enable its own gravity to hold it together at the speed at which it was spinning.

One should then expect fairly large chunks to break off, particularly if the star was still semi-plastic and not yet very strong. These would form orbits at no great distance. But if cooling and consolidation continued, one would expect subsequent broken fragments to decrease in size. By the time these were thrown off the star would have acquired a greater rotational speed and so the smaller fragments would be hurled to greater distances and have larger orbits.

In this account of possible events I have ignored the retarding effect of tides. I think this is justified. For the assumption is that the sun's com- panion was of a mass comparable with that of the sun at the time when it underwent this succession of adventures. On so massive a star the retarding effect of tides would be small in comparison with the accelerating effect of extinctions.

D.6: *After the Planets had been thrown off*
I have mentioned some of the grave objections that must be urged against the original double star theory and have shown how they can be met, without the invention of *ad hoc* hypotheses, but only by inference from Symmetrical Impermanence. But there remains one further objection. It is quite as serious as those already mentioned and is raised by the simple question: Where is the

parent star now ? The double star theory cannot be accepted unless at least this question can find a satisfactory answer. A body of the size of the sun cannot be hidden.

The only answer that I have come across is that the sun's companion, after throwing off the planets, was itself propelled far away into outer space. It has even been suggested that it can still be identified as a faint and distant star. But I have failed to find any attempt to account for the propelling force.

This could not have been generated by the helium synthesis nor by the synthesis of the heavier elements. However violent the happenings may have been during these processes the forces acted from within the star; they could not have exerted a force on it from without.

Centrifugal force is equally unable to account for the removal of the star. It is, after all, a matter of elementary mechanics that centrifugal force acts *from* and not *on* the centre of gravity of a rotating body. Centrifugal force cannot displace the centre of gravity.

It is thus evident that the double star theory, as so far presented, is quite unsatisfactory. It accounts for the materials of which the planets are formed and this is its great merit. But it leads one to expect a star of comparable size to that of the sun within the solar system and does not account for our failure to observe it. Advocates of the theory assert that this star was removed but do not explain the removal. To do so they would have to invent an additional *ad hoc* hypothesis. This would have to show how a force was exerted on the sun's companion from somewhere, how this force carried the star off to a distant place, and how it did so without interfering with the sun. The force would have to be shown to have operated after the planets had been thrown off. To explain the removal of the sun's companion we have to postulate something that acted from without, like a billiard cue.

Given sufficient ingenuity one would, no doubt, be able to invent some possible hypothesis. But even that would not save the double star theory. It would still need to be explained why the planets were left behind. It must not be forgotten that, according to the double star theory, the planets would at the time not have been circling the sun but have had their orbits around the star from which they had been torn. They would have been within the gravitational domain of this star. Where the star went, its planets would go too. This, again, is a matter of simple mechanics. So further ingenuity would be required in order to invent a process that would, as it were, pick the planets off their parent star and transfer them to the sun before the parent star was propelled away.

As hitherto presented, the double star theory is therefore quite unacceptable. And yet it is the only one that explains satisfactorily why the earth and other

229

planets contain all the elements. May it be true after all, in spite of its apparent absurdities?

D.7: *Jupiter in a New Light*
Let me list the modifications to the original double star theory that I have suggested above:

The sun's companion was not larger than the sun but smaller.

Exhaustion of the hydrogen in the sun's companion was not the consequence of a sudden rise in consumption but of the lack of replenishment that one would infer from Symmetrical Impermanence if the sun's companion were the smaller partner and could not compete successfully with the sun for hydrogen.

Synthesis of the heavier elements did not occur during a catastrophic rise in activity but more slowly and as a result of the close packing of nuclei after collapse of the star.

The parent star consisted of a solid core surrounded by a very thick envelope of gas. The planets were not formed from this envelope but from the solid core. In their passage through the envelope they took some of this with them. Therewith they acquired an atmosphere, though not relatively as voluminous a one as that of the parent star. Their specific weights were lower than that of the parent.

No planets broke off until the sun's companion had lost considerable mass by extinction and had therefore acquired a greatly increased angular velocity.

This history is admittedly hypothetical and will have to be studied critically by experts not only in astrophysics but also in other branches of science. I have already said that I am presenting here a programme for research and not a conclusion that has already been fully tested.

Among the questions that have to be put to specialists are: Would synthesis of the heavier elements occur if the sun's companion collapsed comparatively slowly, as postulated in my revised theory?

Would the synthesized substances acquire a density similar to that of the earth?

What is the quantitative relation between the tensile strength of the substance thrown off and the distance to which it would be flung? Could planets have acquired sufficient angular velocity to throw off moons and have been subsequently slowed by tides down to their present angular velocity?

230

Or could the moons too have been thrown off by the parent star and subsequently become attached to this or that planet?

For the moment I can only hope that answers to these and some other questions will prove to be consistent with the revised double star theory, quantitatively as well as qualitatively. On the assumption that it is so, some further conclusions follow from Symmetrical Impermanence. These are not hypotheses any longer; they are inferences from continuous extinction.

Being smaller than the sun the companion star would continue to lose mass by extinction after it had thrown off the planets. The smaller it got, moreover, the more it would fail in competition with the sun to obtain adequate replenishment by capturing hydrogen from outer space. The discrepancy between the masses of the two stars would thus become greater and greater as time went on. The companion star would dwindle in size and the smaller it became the less hydrogen it would capture until it captured virtually none and its size would be wholly a function of the half-life of matter. Given time enough one would infer that the companion star would dwindle to a mass that was but a small fraction of the mass of the sun.

For a while the shrinking companion would continue to be the centre of the circling planets. But it would also be becoming more and more itself a planet of the sun. As its mass decreased, the orbits of its planets would also become less and less like ellipses; the sun would exercise an increasingly pronounced distortion. It would, as it were, begin to lure the planets away from their parent star. When this became very small it would lose all hold on its offspring. The planets would then edge bit by bit towards orbits with the sun as centre and would finally move in these. The effect of the companion star would be only to prevent the orbits from being perfect ellipses. But even this effect would diminish with time.

At a certain date the mass of the parent of the planets would be about one-thousandth that of the sun. This, I suggest, is the date at which we have now arrived. The sun's companion was not propelled away. It is still in the solar system and we can see it on any clear night. It is a large golden sphere and is called Jupiter.

If this theory is correct it throws some light on the constitution of Jupiter and the other big planets. Their density is much lower than that of the earth. But if all the planets are offspring of Jupiter this latter must have a solid core consisting of much the same substances as the earth and with a density that does not differ greatly from that of the earth. From that it has to be concluded that Jupiter is surrounded by a very thick envelope of lighter material, which may be liquid or gaseous and may consist in its outer layers largely of hydrogen.

On my theory of the origin of the planetary system one should expect fragments thrown off from the parent star to carry a relatively small amount of liquid or gaseous substance from that star away with them; for while being buried through the atmosphere of the parent star they would leave most of it behind. But some atmosphere would cling to the fragments by virtue of their gravitational field, though it would be but a small fraction of their mass.

One should expect this fraction to be greater the greater the mass of the fragment. The other large planets besides Jupiter should, therefore, be expected to have a much thicker liquid and gaseous envelope than the earth and, therefore, to have a smaller density. One should expect Jupiter itself to have the thickest envelope and lowest density of all. Observation confirms this expectation (with the exception of Saturn).

Appendix E

The Shrinking Earth

E.1: *Further Means of Estimating the Half-Life of Matter*
These appendices are, as I have said already, to be regarded as research programmes. One of the subjects on which research is needed is the half-life of matter. A reason has been given in Appendix C why this cannot be much greater than 4×10^8 years and why it is probably not much less than 3×10^8 years. In this appendix further means for estimating the half-life will be given. They contain some uncertainties, which the future may remove ; and in view of these more importance must here be given to the method than to the estimated number of years to which it leads. It will be shown that more accurate knowledge than we have at present is likely to lead to an estimated half-life not greater, and perhaps substantially less, than 4×10^8 years.

It has been suggested in Appendix D that the elements of which the earth is composed were formed out of hydrogen in Jupiter at a time when that planet was much more massive than it is now. It is this hypothesis that can help with an estimate of the half-life of matter.

There is evidence that some of the earth's sedimentary rocks were laid down 2×10^9 years ago. It may, for all I know, be impossible to decide whether this happened while the earth still formed a part of the substance of Jupiter or after it had become a planet. But the former view would seem difficult to maintain and, in the absence of evidence to the contrary, most of us, I feel sure, would assume that the first sedimentary rocks on the earth were formed after this planet had been hurled into independent existence. When this happened the speed of rotation of Jupiter must have been such that centrifugal force substantially exceeded gravity.

The most recent estimate for the age of radium-bearing rocks is greater than that of sedimentary rocks, namely 2.6×10^9 years. This is the time that, according to the revised double star theory, has elapsed since the heavy elements were synthesized in Jupiter. When it happened the speed of rotation of Jupiter must have been such that gravity substantially exceeded centrifugal force.

These figures show that the speed of rotation of Jupiter must have risen greatly during less than 6×10^8 years. At the beginning of this interval it was such that gravity produced the pressure needed for the heavy elements to form. At the end of the period it was so much greater that the effect of gravity was more than cancelled; there was a surplus of centrifugal force sufficient to overcome the cohesive strength of the material.

It has been shown in Appendix D that only one reason can be suggested for a large increase in the speed of rotation of a solid star not subjected to an external couple, namely loss of mass by extinction. Hence the time interval between the birth of the heavy elements and the birth of the earth could help with an estimate of the half-life of matter. One would have to know the centripetal force needed to form the heavy elements and the centrifugal force needed to eject the planets. The sum of these forces would be a measure of the increase in speed of rotation between the two events. This, in turn, would be a measure of the loss of mass by extinction. It ought to be possible to arrive at rough estimates of all the quantities that enter into the calculation. It seems to me that a threefold increase in speed of rotation during 6×10^8 years is of the right order of magnitude; and this corresponds to a half-life of 4×10^8 years.

Another clue is given by the present ratio of the mass of Jupiter to that of the sun.

According to the revised double-star theory Jupiter has never been quite as massive as the sun, but it seems to me that it must at one time have been nearly as massive, for I doubt whether a double star can form at all unless the masses of both stars are about equal during growth. If one of them were much less massive than the other, one would expect it to be absorbed by its larger neighbour during an early stage of its development. Hence each partner of a binary star system probably begins to form at about the same time as the other partner and to grow at about the same rate until both have become so massive that income by capture of surrounding matter and loss by extinction roughly balance. A small difference in the income and loss account would then be critical.

This would be the moment when a difference between the two masses would decide the subsequent course of events. Incapable of competing successfully with its larger partner for surrounding matter, the smaller star would find itself after a certain time on the wrong side of the balance sheet; it would be losing mass. The net loss rate would be small at first, for it would be but a small difference between the rate of extinctions and the slightly lower rate of captures. But as the difference between the masses of the two neighbouring stars increased, the rate at which the smaller one replenished its substance by capture would also decrease; the discrepancy between the two masses would become ever more accentuated. While the larger star would continue to balance its accounts and might even continue to grow, the smaller one would in time become so small that its income from capture became negligible and its future size would be wholly determined by the half-life of matter.

234

If the above reasoning is sound, the sun's mass has not changed by a large factor since it reached its maximum value a certain number of thousands of millions of years ago and Jupiter had the same mass at one time; the present mass of this planet is then one-thousandth of the maximum value that it ever reached. The reduction began slowly, while loss was being partly compensated by capture, but became more rapid when capture practically ceased. The curve connecting the mass of Jupiter with time must then have had the general shape shown in Fig. 6.

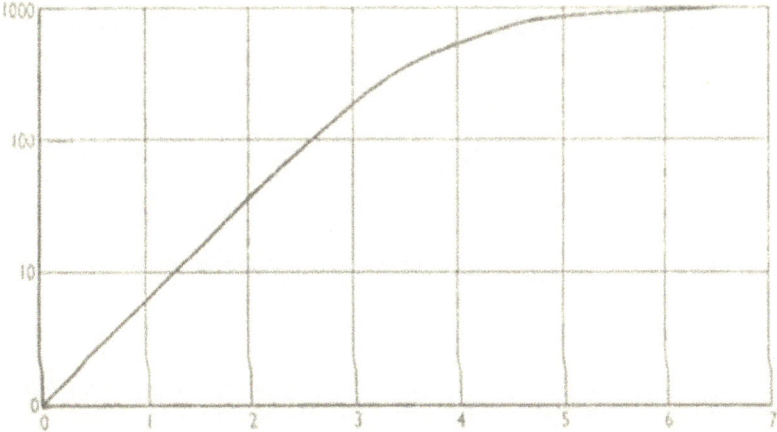

X 10^9 YEARS AGO

Fig. 6. *Relation between mass (vertical scale) and time (horizontal scale) for Jupiter, if the half-life of matter is 4 x 10^8 years. The figures on the vertical scale give the past mass of Jupiter as a factor of the present mass.*

The curve has been drawn for a half-life of matter of 4 x 10^8 years. Other curves could have been drawn to the left for shorter and to the right for longer half-lives. But they would all have to begin at the past time when Jupiter and the sun were both about equal and when both began their existence as complete stars. Estimates seem to vary as to when this was, but I cannot recollect one greater than 7x10^9 years ago. It is difficult to draw a convincing curve that shows the mass of Jupiter to have been 1000 times its present value 7 x 10^9 years ago and that also corresponds to a half-life greater than 4 x 10^8 years. So I should like to adopt this estimate provisionally while hazarding the guess that more reliable data will lead to a rather shorter estimate.

E.2: *The Diminishing Value of g*
Let the mass of the earth t years ago be m_o and its present mass m. Let the radius of the earth that corresponds to m_o and m be, respectively, r_o and r, and let the half-life of matter be T_m. It can be shown easily that the following relations hold for small values of t:

235

$$m_o / m = 1 + 0.69t / T_m$$
$$r_o / r = 1 + 0.23t / T_m$$

From these equations, and if T_m is taken at 4×10^8 years, the following conclusions are reached:

(a) The value of the gravitational constant, g now expressed in metres per second square as 9.80665, had more nearly the value 9.80667 in Newton's time and will have the value 9.80663 in the year 2250.

(b) The radius of the earth is diminishing by nearly 4 mm. per year.

(c) During the geological period known as the Cambrian the earth had about two-and-a-half times its present mass and its radius was about 13.5 per cent. greater than now.

(d) During the time, two thousand million years ago, when the first known sedimentary rocks were laid down the mass of the earth had thirty-two times its present value and the radius was more than three times as great as it is now.

(e) At the time when these sedimentary rocks were laid down water boiled at 140°C. Its great weight must have caused it to have an enormous scouring action on rocks. Sedimentation must have been a much more rapid process than it is now.

(f) If it were not for the slowing down effect of the tides, the continuous decrease in the earth's radius of gyration would lead to a continuous increase in its angular velocity. It follows that tides dissipate more energy than has hitherto been supposed.

The astronomical, geological and biological consequences of this slow but unceasing reduction in the earth's mass and volume must be significant. If my theory that the half-life of matter is finite is true, it must therefore provide extensive and rewarding fields for new research by experts in these three disciplines. This appendix is not a suitable place for considering in detail the nature of this research, but it is worthwhile to consider some samples of the problems that continuous extinction raises in each of these disciplines.

The astronomical field seems to be the least promising of the three, for the astronomical changes that would result from shrinking of the earth are likely to be the least easy to detect.

One such change would be the earth's hold on the moon, which must be steadily loosening as g decreases. It has already been mentioned in Appendix C. Another would be the increase in angular velocity that must accompany reduction in the earth's radius of gyration. As just mentioned, this speeding-up of the earth's rotation must go some way towards counter-acting the slowing-down effect of the tides. But I cannot think of any astronomical observations precise enough to detect the predicted changes in g and in the radius of gyration.

236

These changes would affect the exact times when eclipses occur. One can calculate these times very accurately for past eclipses on the assumption that the earth's mass was no greater then that it is now and could then compare the result with historical records. If observations thousands of years ago had been as precise as they are today, one might be able to discover from a discrepancy between the calculated and the observed moments whether the earth's mass has changed or not.

But to be sure one would also have to disentangle the two effects that influence the length of the day: tides, which tend to make the day grow longer, and contraction, which tends to make it grow shorter. For this one would have to know accurately how much energy is dissipated by tides and this is, in fact, hardly possible to estimate. The conclusion is, I am afraid, that precise astronomical observation and calculations are too recent and knowledge of some of the relevant quantities too incomplete, for the gradual shrinking of the earth to have an observable effect in astronomy.

E.3: *Geological Implications of Shrinking*
In geology the position may be more hopeful. The finite half-life of matter may, for instance, succeed in explaining the formation of mountain ranges. The only explanation of these that has so far seemed at all possible is that the ranges are the consequence of shrinking of the earth. But it has hitherto proved impossible to find a satisfactory cause of the shrinking.

At one time it was believed that this cause was thermal. In remote geological times, it was thought, the earth's core was much hotter than it is now. As it cooled it contracted and the more rigid crust ceased to be a tight fit. Surrounding the contracting core loosely it came under the influence of strong shear forces, which caused it to pucker into mountains and depressions. 'Like the skin of an apple', it used to be said.

It was always difficult to understand how mere thermal causes could suffice to explain the contraction; for the change in the volume of a solid substance with change of temperature is not great and puckering seems to occur only if the change of volume is rather great. It must be remembered that what has to be accounted for is not one single pucker but a succession of many such at irregular intervals of time.

The earliest ranges have long since been weathered down by the action of water. Their broken and distorted remains have sometimes been raised into new ranges by a subsequent upheaval. Mountains arose in Great Britain, Norway and Greenland during the Silurian period, 350 million years ago. They have been largely obliterated and have given place to later mountains, often on the same sites. The more recent Appalachian Mountains were formed in the days when the great reptiles lived and crawled, we are often told, in swamps. These mountains were followed still later by the Sierra Nevada, the

Rocky Mountains and the Andes, which arose in the Cretaceous period. The youngest arrivals are the Alps, the formation of which coincided with some of the higher mammals. A considerable amount of continuous or repeated shrinking is needed to account for so much.

Thermal shrinking can certainly not do so and the theory that it did has long since been abandoned by geologists. The earth has cooled but little, if at all, during geological times, as is demonstrated by sedimentary rocks. These have been precipitated out of the sea and some of them were formed, as has been mentioned already, more than two thousand million years ago. So apparently the sea existed at that remote time. The average temperature of the surface of the earth must therefore have been below the boiling point of water during the geological ages in which it is known that the mountain ranges were formed. It is still above the freezing point of water. If the average temperature of the earth has differed at all from one geological upheaval to the next, the difference must have been negligibly small.

Such considerations have led geologists to seek some other explanation for mountain ranges. There are at least four but it would not be helpful to discuss them here in detail. None can be said to hold the field. The conclusion at which one arrives is that it is almost, if not quite, impossible to account for the puckering of the earth's surface if there has been no shrinking and it is, similarly, almost, if not quite, impossible to account for any sufficiently pronounced shrinking so long as one assumes the half-life of matter to be infinite. The best that can be said for the various rival theories about the origin of mountain ranges is that they are valiant attempts to explain what still seems to be inexplicable.

In these circumstances it is worthwhile to consider whether the puckering of the earth's surface could be the consequence of the kind of shrinking that results from the continuous extinction of matter. To prove that this was indeed so would be to provide somewhat sensational support for Symmetrical Impermanence. But it can only be done if a rather serious difficulty can first be surmounted. Its nature must claim our attention for a moment.

The extinction of an elementary component of the material universe is accompanied, it has to be remembered, by the simultaneous extinction of an equivalent volume of space. Hence we have to face the following question: when an elementary component becomes extinct, does the space occupied by this component become extinct as well? If it does, the extinction does not leave a void. The matter that has been adjacent to what has disappeared is not adjacent to the place where that was, for the place, too, has disappeared. The compacting of matter is, on such an interpretation, exactly the same after a disappearance as it was before.

If continuous extinction has to be interpreted in this way, the extinction of matter within the earth would change its size without changing its density or its shape. If nothing happened to the earth except continuous extinction, the earth would, at any particular moment, be an exact reduced scale model of what it had been before. Shrinking of this kind would not impose any shear forces and could not lead to any puckering. If the extinction of matter is accompanied by extinction of the space previously occupied by that matter, the origin of mountain ranges must remain as inexplicable as ever it was.

But I am not sure that this is the correct interpretation of continuous extinction. The relation between matter and space is still very obscure and no one can honestly say that he has no difficulty in appreciating the meaning of the relativistic view that space is a constituent of the material universe. For most of the time we tend to ignore this conclusion and continue to regard space as a container. Much clarification of thought has still to be done before any of us can hope to discuss adequately the relation between space and matter. At present we can ask the question whether it is true that space and matter originate together and become extinct together. But we cannot hope to answer the question until we understand its meaning much better than we do today.

Having uttered this warning against both hasty acceptance and hasty rejection of any conclusion about the relation between space and matter, I venture to draw attention to a conclusion that will be put forward tentatively in Appendix H. This is that, when an elementary component of the material universe becomes extinct, the space that becomes extinct, too, is not that occupied by the component, but surrounding space. It will be shown that it is more consistent with the relativist notion of a differentiated space to assume that a wave of contracting space travels outwards from the site of an extinction without causing disappearance of the site.

If this conclusion does, in fact, follow from a correct appreciation of the relation between space and matter, a void is left behind whenever an elementary component of the material universe becomes extinct. As extinctions are occurring continuously in all substances, such voids are also occurring continuously. Their subsequent fate must depend on whether or not the substance is subjected to a compressive force. If it is not, the voids remain and the substance loses mass without losing volume. Its density decreases with time and would be halved in time T_m But if there is a compressive force, the voids are squeezed up. The loss of mass is accompanied by a loss of volume and the density is not changed.

These conclusions can be applied to the shrinking earth. The consequence of continuous extinctions is that the core, which is subjected to the weight of the superimposed layers of the earth and the atmosphere above the earth's surface, loses volume as well as mass, while in the first instance the outer crust loses mass only and does not lose volume until gravitational forces and the weight of the air above compress it. Hence the crust ceases to be a tight fit and the puckering that geologists attribute to contraction takes place.

I have purposely not elaborated the above theory and I put it forward with considerable diffidence. I should not do so at all were it not that the formation of mountain ranges, like many other scientific mysteries, calls for the collaboration of specialists in many fields of study and the hypothesis of continuous extinction does at least offer a prospect, be it but a faint one, that this mystery will be solved by a joint effort of geologists, mathematical physicists, nuclear physicists and relativists. There are occasions when an author ought to withhold a new theory until he has tested it from every point of view. But when the testing calls for the combined efforts of many, his proper course is to offer the theory to the world while it is still in a tentative state.

Let me add that, according to this theory, there is no filling up of voids when there is no compression. There is none on the surface of the moon or on Mars, which have no atmosphere to weigh on the solid surface. One should therefore expect the surfaces of these bodies to have become more and more spongy with time. Mattel there must be in a condition unknown and unreproducible on earth - solid, yet with a very low specific gravity. If my theory is correct, comets must consist of similar substance, for they must be the degenerate remnants of larger chips thrown off Jupiter.

E.4: *Continental Drift*
At the beginning of this century there was much discussion of a theory that was called, after its originator, the Wegener Hypothesis. According to this the earth's geography has undergone radical changes during geological times. The Eastern and Western continents are supposed, for instance, to have been joined together into a single huge landmass. If it was so, some cataclysm must have broken the mass into two pieces, of which one is now the two Americas and the other the combined continents of Europe, Africa and Asia. After their severance the two pieces must have drifted apart, leaving the Atlantic Ocean between them.

At least four arguments have been urged in favour of this hypothesis and one against it. Here there is neither the space nor the need for their detailed presentation. As I have said already the purpose of these appendices is to stimulate research and not to provide an exhaustive study of each of the many subjects on which Symmetrical Impermanence seems to throw a new light. Those who are qualified to explore the effect of continuous extinction on the earth's geology and geography must be acquainted with the extensive literature on the Wegener Hypothesis and the recent revival of interest in it. So a brief enumeration of the arguments for and against it will suffice.

The Wegener Hypothesis has never been presented as an inference from
240

accepted basic facts and principles. It is an *ad hoc* hypothesis devised to explain certain observations for which no other explanations have been found. Its strength is its considerable explanatory power. Its weakness is that it itself calls for an explanation and this has not been forthcoming. The first set of observations explained by the Wegener Hypothesis is biological. Certain species of animal, for instance, occur on both sides of the Atlantic that are nearly related and must have had a common ancestry in times that are recent by the geological time scale. Descendants of these common ancestors must at some time have crossed from the one land mass to the other. But they would have had to cross the sea to do so if the masses had always been separated, as they are now.

One might suggest that the masses were joined only at their Northern ends, where they are now nearer to each other than anywhere else. But if so, descendants of the common ancestors would have travelled a very long way; for existing nearly related species are to be found, respectively, in the Southern ends of South America and South Africa. If the present species have descended from emigrants successive generations would, during their wanderings, have been adapted to the variety of climate that occurs between arctic, moderate and equatorial latitudes. But there is no evidence for such an evolutionary history. The hypothesis that is most consistent with the biological observations is that the Southern tips of South America and Africa were joined together at the time when the common ancestors of the present species lived.

The second set of observations is purely geological. The East coast of the Western continents and the West coast of the Eastern ones fit together like pieces in a jigsaw puzzle. The fit between the edges of the continental shelves of the two landmasses is even closer than that between the coastlines that appear on an ordinary map and that show only the limit of the land that is above sea level. There is, moreover, a ridge running from North to South in the Atlantic such as one might expect if there was a continental break at that place.

It has also been shown that the remaining landmasses, such as Australia and Greenland, can, with a little ingenuity and rotation, be fitted together. This has led to the suggestion that the whole of the earth's land surface was continuous at one time. Sundry bits are supposed to have broken off occasionally. But when they did they can never have stayed close to each other for long. Something must always have caused them to drift apart.

The third set of observations is from mineral deposits. Gold is found near the East coast of South America and in the corresponding part of the West coast of Africa. The gold is alluvial and such as one should expect in the lower reaches of a river. It is so on one side of the Atlantic but not on the other. Its presence there is therefore difficult to explain. But it would be quite explicable

241

if the gold in each country had been deposited by the same river when the continents were joined together.

The fourth set of observations is of certain magnetic rocks. Particles of iron ore in these are little magnets, each with a North and a South pole. In a given sample of rock all the magnets point the same way.

The best explanation of this uniform direction that has been offered is that the particles aligned themselves to the direction of the earth's magnetic field at the time when the magnetism occurred. But the alignment today is often far from the direction of the poles. One explanation of this would be that the direction of the earth's magnetic axis has changed greatly in past ages. But there is no other evidence for this, besides which the geographical distribution of the rocks in question and their ages are hardly consistent with this explanation.

The explanation that does seem to fit many, though not all, the facts is that the particles oriented themselves to the magnetic poles when they were in one latitude and longitude and have been carried since on the landmass that they occupy into other latitudes and longitudes.

The objection to the theory of continental drift is that no reason can be found why the landmasses should drift apart. Even supposing something did cause a crack to occur right across the earth's crust, from one end to the other, why should the two pieces not stay near each other? It is true that the earth's crust floats on a more or less plastic mass. A force applied horizontally to a portion of the crust would overcome the stickiness of the plastic material and cause movement. One could argue that the force need not be very great if the movement is very slow. But for even slow movement there must be a force of a significant magnitude and friction has to be overcome; potential energy has to be converted into frictional energy, into heat. So long as no force at all of the required magnitude and direction and no suitable source of potential energy are accounted for, the Wegener Hypothesis has to be abandoned.

It is here that the hypothesis of the continuous extinction of matter promises to help. It can explain both the breaking off of landmasses and their subsequent movement.

Let us consider an extensive continent floating on a plastic mass. If continuous extinction causes a reduction in the earth's radius such as is given by the inferred half-life of matter there will be a tendency for the more rigid continent to retain the previous radius while the plastic substratum acquires the new one. When this happens the continent will form a cantilever. If it is too rigid to bend it will break. At the place where the break occurs one should expect the crust to be firmly embedded in the plastic mass. Some of the crust would be left there, as the stump and roots of a tree are left in the ground

242

when a gale blows the tree over. The ridge that stretches along the Atlantic may well be the 'root' of attachment of the earlier continent, of which two pieces now form, respectively, the Western and Eastern worlds.

For the reason why the pieces should drift apart one has to turn to one of the theories that have been brought forward in the hope of explaining the formation of mountain ranges. It is as follows:

Radio-active materials within the earth are a source of its heat. It is the disintegration of the atoms in these materials that has enabled the earth's temperature to remain approximately constant during thousands of millions of years. The heat is radiated away at about the same rate as it is produced. But the rate of radiation cannot be the same for all parts of the earth's surface. The continents form a blanket and keep the heat in while the oceans allow it to escape readily. From this one must assume that the earth is hotter deep below the continents than it is below the oceans.

The higher temperature must be accompanied by thermal expansion. This will produce an upwards thrust under the continents, which will be opposed by their weight. But when a crack has occurred there is nothing to restrain the plastic mass below the site of the crack from expanding and raising the edges of the crust at each side of the crack. These edges will then be further from the centre of the earth than the outer edges. Thus two inclined planes are formed and the separating continents slide down them in opposite directions.

The force that causes continental drift is, according to Symmetrical Impermanence, gravity. The original source of the energy that overcomes friction is radio-activity within the earth. This is converted into heat and thereafter, to a small extent, into the potential energy of position that is measured by the raising of the ground level where a crack occurs.

Thus a person who had no other reason to postulate continental drift would, nevertheless, postulate it if he accepted Symmetrical Impermanence. This, in other words, transforms the Wegener Hypothesis from the category of *ad hoc* hypothesis into that of inference.

E.5: *Biological Implications of Shrinking*
Biological considerations must claim our attention next. Here there is a more immediate prospect that a study of the consequences of a shrinking earth will be rewarding. For a large change in the mass of the earth could hardly fail to have its pronounced effect on the things that live on it. If the half-life of matter is of the order of magnitude that I have estimated, conditions must have been significantly different in those remote days when those plants and animals were alive whose fossils have been discovered. One should expect the difference to be reflected in their structure and their habitats.

243

According to the now accepted view that the half-life of matter is infinite, there is no reason to suppose that physical conditions on earth have changed very significantly during the past two thousand million years. I have already pointed out that the temperature must have been about the same in the remote past as now. There must have been sea and land and the same atmosphere. Science fiction may speak of swamps and twilight and a primordial slime, but geology does not seem to justify such flights of imagination. Nothing is known against the conclusion that the same plants and animals that live today, including man, would have found the earth a congenial place in those distant times. Why then, one is led to ask, did they not occur for so many hundreds of millions of years? Was the earth really, it has to be asked, hospitably ready so long ago for the forms that exist today and did they really wait all that while before they accepted the invitation? If the earth had the same climate and other physical conditions so long ago, why the grand evolutionary cavalcade, in which species evolve, have their brief era of dominance, become extinct and are succeeded by others? Why did those species that occur now not evolve earlier? Why did the early species not survive to this day?

One possible answer is that evolution is a slow process and that during those past hundreds of millions of years plant and animals have been adapting themselves ever better to conditions on this planet. But the answer is hardly tenable. We know that evolution is not as slow as all that. Pronounced evolutionary changes come about during periods of time that are quite short by the geologist's time scale. The complex history of mammalian evolution was completed in only a couple of hundred million years. At this rate one should expect that evolution from amoeba to mollusc would have been completed at least eighteen hundred million years ago. But the first known molluscs seem to date from Cambrian times, only five hundred million years ago. Fifteen hundred million years for evolution of the first thing that could leave fossil remains is too long. It does not equate with modern genetic knowledge.

We also know that the species that are now extinct must have been well adapted to their surroundings, for they did survive for many generations. On a rational interpretation of evolution there can have been but little scope for improvement. Yet no species that lived in Cambrian times has descendants today that resemble their ancestors sufficiently to be classed as the same species.

A possible answer to the question why every early species has become extinct is that there was too much competition with more successful rivals. According to this view every kind of plant and animal that ever lived on earth would have living descendants like itself today were it not that it was either preyed upon or starved to the point of extinction by more competent later arrivals. To say this is to say that adaptation to the physical circumstances of the inanimate

244

environment was perfected a long while ago and that since then adaptation has always only been to the competitive circumstances introduced by living things.

This answer might explain the occasional extermination of a few species but not, it seems to me, that of every species that existed a few hundred million years ago. It is true enough that Nature, as the poet said, is red in tooth and claw; the struggle for survival is indeed fierce. But the opposed forces are closely balanced. The conflicts that we observe today rarely result in the complete elimination of one of the warring species. What more often happens is that the one gains no more than a relative ascendancy over the other. The world is, moreover, not so densely populated that the various species are in continuous conflict. Many survive by the simple process of avoiding those stronger than themselves. Animals that are preyed upon remorselessly go on having descendants for many millions of years. Why should the day have come for every one of them when there could be no more descendants of the same Form?

The answer that physical conditions have been steadily changing seems to me to provide a far more satisfactory explanation of the evolutionary trend than the answer that the sole cause of change has been competition. But no other reason for a change in physical conditions seems to have been found except the change in the mass of the earth that I am claiming. This change means a corresponding change in the value of g and must have very pronounced and varied effects. In this research programme I do not pretend to give an exhaustive list of them, but only to mention some samples .in the hope that others may further explore the field. I feel sure that it will prove a rewarding one.

If the half-life of matter is 4×10^8 years, the weight of everything was more than three times its present weight two thousand million years ago. rhis must have caused the earth to be a very different place to the congenial earth that we know.

The air pressed on the earth with a force of over 3 kg per square cm. The atmosphere was squeezed into a much thinner, if denser, blanket than our present atmosphere. There will have been but little evaporation of water into the dense air. Clouds must have been rare. Far from the twilight postulated by the romancers, the land and the sea must have been under a fierce glare. When clouds did form and shed their rain, each heavy raindrop must have fallen with a great speed and reached the ground with a truly destructive impact.

At the high density of the air small local differences in temperature must have caused large displacements of air, so that average wind velocities must have been very high. At the same time even a low velocity, such as would be a gentle breeze today, must have had the force of a hurricane when the air was so much denser. The gales that swept the land must have been devastating and will have carried sand and pebbles, even great boulders, along with them.

When conditions on land were so violent, the sea must have been the only safe place for living things. If any vegetation at all could maintain .itself on land, it must have been of the kind that clings to rocks like some of the present lichens. Anything that protruded a frond above the rock face would have been torn off by the raging tempests.

Before there were plants to put oxygen into the air and to provide food there could be no animals and as there could be no land plants until g had fallen to a level at which the earth had a tolerable climate, one should not expect land animals to have appeared until comparatively late in geological times. Even sea animals must have kept themselves well out from land. For only the tiniest of creatures could have survived the violence with which waves must have broken on the seashore in those times.

When g had several times its present value, a branched tree was an impossibility. For no kind of wood would have been strong enough to support those cantilevers that we call branches. Stalks could not have borne the weight of fruit. Even leaves would have been a burden. If a plant could find a place to grow sheltered from the gale, it would yet have to have such a structure that its leaves would not drop off by their own weight.

Similar considerations apply to the early animals. No bull could have been strong enough to support its cantilever-like head. Bones and muscles could not have supported the tissues of any large animal. So one need not be surprised that, right up to the carboniferous period, there were no large animals. It was only then, when, by my estimate, g had not much more than twice its present value, that the amphibians began to evolve.

When g is great it is obviously easier to crawl than to walk. Hence the great saurians, which lived in the days when our coal deposits were being formed, were eminently adapted to terrestrial conditions of the time. It may also well be that they did really live in swamps. For they could then keep their heavy bodies submerged so that some of the weight would be borne by the water. As g decreased, their descendants will have raised more and more of their bodies above the surface. But nimble four-footed creatures had to wait until the weight of the mammalian body had dropped to a value at which slender stilt-like legs can support it; and then the first such creatures could only be small ones.

So far I have only discussed those conditions that, from our present point of view, seem to have made the earth increasingly hospitable. I have spoken of the conditions that made later arrivals possible. But did the decreasing value

246

of g not perhaps make the earth less hospitable to its previous inhabitants? Was there no change of the kind that would make the earth impossible for them?

There must have been such changes, too, and a rewarding field of research would be to discover what they were. I shall be content to mention one obvious one. It is the decreasing density of the air. The heart, the blood and the lungs of the great saurians must have been adapted to the air pressure of their day. As this pressure decreased, they must have found themselves starved of oxygen. It may be that the descendants of us, who are alive today, will become extinct for the same reason. According to the hypothesis of continuous extinction, the time must come eventually when the earth will have so little mass that, like the moon now, it will no longer retain its atmosphere. But long before then the air will have become so rarified that creatures adapted to a pressure of one kg per square cm will not be able to get enough air to support life. Neither a violent cataclysm, nor cold, nor heat, but lack of oxygen seems most likely to bring the end about.

If gases and solids have been heavier in the past than they are now, so, of course, have liquids. When g was much greater than it is now, sap could not have risen in a tree. Nor could any animal have had a heart powerful enough to pump the heavy blood to a level much above that of the heart. The shape of a lizard was then better adapted to environment than the shape of a giraffe. While the heavier air could support the heavy body of a flying animal, monkeys could not readily have swung from branch to branch of trees.

In particular, one may surmise that man's erect posture and the enormous development of his brain had to wait until g had fallen to an appropriate value. A creature that held its head well above the ground and whose brain needed so copious a supply of blood could not have survived in a world where blood weighed much more than it does now.

Appendix F

The Inverse Square Law

As is well known the inverse square law applies equally to electric charges, to magnet poles and to masses. It is often formulated more or less in these terms: 'The force exerted by an electric charge on another electric charge is inversely proportional to the square of the distance between the centres of gravity of the charges; this statement is equally true if one replaces the words "electric charge" by "magnet pole" or by "mass".' One would hardly expect so simple and familiar a law to allow scope for differences of opinion, and yet one meets various interpretations of its meaning and status. They are rarely stated explicitly, but are implicit in the way people talk about the law. I propose to ventilate this subject here because I believe that misconceptions about the inverse square law are symptomatic of a general misconception that is forming an obstacle to the unification of physical science. It is the misconception mentioned in Part I of this book, the notion that the laws of physics are specific and selective, that they are of the statute book kind. I select the inverse square law, not only because it is a typical example but also because I propose to show in Appendices G and H that a proper appreciation of this law offers a prospect of better insight into the relation between space and matter.

Many would, I feel sure, endorse every word of the following statement about the inverse square law:

'There are really three distinct inverse square laws. One of them defines the nature of electric charge, one the nature of magnet poles, and the third the nature of mass.

'Each of these three laws has become known by experiment and observation and could not have become known by any other means. These essential tools of the scientist had to be applied specifically to each of the three sources of force that are mentioned in the law. Thus the inverse square law for electric charges was discovered by experimenting with such charges; the inverse square law for magnet poles by experimenting with these; the inverse square law for masses by observing the movement of masses under the influence of the forces exerted between them. To deny that the three sets of distinct experiments were necessary is to forget that physics is essentially an experimental science. That the force was found to be the same function of

248

distance in each of the sets of experiments is just one of those scientific facts that cannot, and need not, be explained.'

'If there were a Cosmic Statute Book in which all the laws of physics were recorded, the inverse square law would have to occur there in three different sections. They would be those headed respectively "Behaviour of Electric Charges", "Behaviour of Magnet Poles" and "Behaviour of Masses". There would be no statutory means of ensuring that each of these natural phenomena obeyed the law, unless each of them was specifically mentioned in the formulation of the law.'

'A terrestrial textbook on physics from which the student may learn about these three laws has to be compiled on similar lines. There is, for instance, no means of letting him know how the force between electric charges varies with distance without mentioning electric charges.'

'One could not replace the three laws by one single generalization in which electric charges, magnet poles and masses were not mentioned and that would, nevertheless, allow the student to infer how the force exerted by each varies with distance.'

'Each of these three laws is an independent discovery and none of them could have been found by purely deductive reasoning.'

This account of the inverse square law is in strict conformity with an approach to physics and its teaching that meets with wide approval. Nevertheless, I regard it as misleading and mischievous. By way of explaining why, I can usefully begin with a couple of further statements about the law that I have repeatedly heard quoted on good authority. They seem to have a fair measure of support and both are perfectly consistent with the account of the law that I have given above. I shall give in my own words what I think is a fair representation of the view that I am disputing.

The first statement is as follows:

'The inverse square law for mass has been obtained by observation of the movement of bodies within the solar system. It holds with great precision for the distances between these bodies, but it may, nevertheless, be only a first approximation to the truth. There is no reason why, for instance, a precise formulation should not include a second term, which might not be one of attraction but of repulsion. Instead of being an inverse function of distance, this term might be a direct function. For short distances it would then be negligible, but would predominate at those large distances at which the inverse term became very small. The assumption of such a correction to the inverse square law would provide an *ad hoc* explanation of some cosmological puzzles. One of them is the observed fact that the galaxies are moving away from each other. This may be because, at extra-galactic distances, the term in the corrected inverse square law that represents a repulsion between masses exceeds the term that represents an attraction.'

I have already shown in Chapter 24 that this interpretation of the recession of the galaxies is rendered untenable by simple mechanical considerations and that no satisfactory means have so far been found for avoiding the conclusion that space itself is expanding, originating. I shall show in a moment that the same conclusion results from a proper understanding of the inverse square law.

The second statement is less crude. Its defect is not so much that it is downright incorrect as that it misses something essential. It is as follows:

'The inverse square law for electric charges has been obtained from experiments with charges separated by distances that are conveniently large for laboratory work. It holds with great precision for these distances, but there is no reason why it should also hold for distances too small to be measured in a laboratory. The assumption that the inverse square law fails to apply over the distances that span an atomic nucleus would provide an *ad hoc* explanation for a puzzle in nuclear physics. It is this:'

'According to the inverse square law the force of repulsion between the positive charges in the nucleus would be enormous. The nucleus would not cohere. But the coherence would be explained if the inverse square law does not hold for very short distances.'

The faint suggestion here that, when charges are very close together, Nature has granted them a dispensation so that they need not obey the inverse square law is not helpful. It is axiomatic that the laws of physics are valid everywhere, at all times, and in all circumstances. When appearances are against this, the reason is not that the law has failed to apply, but that it has been wrongly formulated. Many laws are special cases of wider generalizations and should then be so worded that this becomes evident. Let me apply these considerations to the inverse square law and seek statements about it that are not merely first approximations but must be absolutely true, and that do not admit exceptions but have universal validity.

During the search I shall try to reveal the weakness of the various statements that have appeared above between quotation marks. As an aid to clarity and cogency I shall employ an occasional touch of irony and I shall imagine a research worker who carries his faith in the paramount importance of experiment over every other means of gaining knowledge to absurd lengths.

An experimenter seeks to confirm the inverse square law for electric charges. He hopes that, by making his measurements sufficiently precise, he may discover a small correcting term, proving that the law as usually stated is only a first approximation.

Among various pieces of apparatus on his laboratory table there is a metal sphere. He places an electric charge on this. He then suspends a small metal disc in such a way that its deflection is a measure of the force exerted on it. He places a charge on the disc, and brings it, together with its suspension, to various distances from the charged sphere. He expects to find that the force varies inversely with the square of the distance, at least to a first approximation, for is this not what the textbooks say? But he is disappointed. The relation between force and distance is far from confirming the law or even seeming to obey any law at all.

A spring balance has been left on the table from a previous experiment. When this is moved, the force on the metal disc changes. For a moment our experimenter thinks that he has discovered a new law about spring balances. He is about to write in his notebook: 'My experiments show that a spring balance exerts a force on a metal disc in its vicinity.' But he is just saved from doing so when he finds that the force is influenced by many things: by the presence of the table, of the apparatus on it, of his own body. All these things, he then concludes rightly, distort the field.

Realizing this the experimenter decides that the law ought to be worded: '*In an undistorted field* the force exerted by one electric charge on another is, at least to a first approximation, inversely proportional to the square of the distance between the centres of gravity of the charges'.

The experimenter appreciates, of course, that an experiment must be performed under controlled conditions. These require measures to prevent the field from being distorted. It would obviously not be good enough merely to remove the spring balance and other pieces of apparatus. In order to confirm the law as revised, and possibly also to find a correction factor, very careful steps must be taken to ensure an undistorted field. But what sort of a field should it be? Being most assiduous at his experimental work, the experimenter decides to try out a great variety of undistorted fields and to begin with the most uniform shape that he can think of. This is the field that occurs between charged parallel plates. By making them extensive he can eliminate disturbing factors and secure ideally controlled conditions.

He places two large copper plates parallel to each other and separated by a short distance. He puts an electric charge on the plates and moves a small test charge into various positions in the space between the plates. Again, he fails to confirm the inverse square law, even approximately. Instead of being inversely proportional to the square of the distance of the test charge from either plate, the force is found to be independent of the distance.

The experimenter does, however, discover something by this experiment. When he varies the charge on the plates he finds that the force varies in proportion. This proves to him that the force on the test charge is proportional to the flux density. As flux density is defined as the property of the electrostatic field by virtue of which the field exerts a force on an electric charge in it, this discovery is true by definition. It is one of those tautologies that are so helpful in scientific work. Being true, it is a discovery that one may expect also to confirm by experiment.

Failure with the parallel plates suggests to the experimenter that the field between these is probably too uniform. He experiments with many other shapes of undistorted field. But the long story of trial and error need not concern us. Among the shapes of field on our experimenter's list is a uniformly *divergent* one, and eventually he comes to it. He finds it a little difficult to construct such a field, but eventually hits on a good way. He obtains two concentric spheres, each of a conducting material. He decides to place a charge on one of them, which will induce an equal and opposite charge on the other. The Faraday tubes of force between the surfaces of the spheres will then be straight lines all diverging from the centre of the spheres, which is another way of saying that the field will be uniformly divergent.

The experimenter arranges to introduce a small test charge into this field and devises ingenious means of moving this about and measuring the force on it. Provided the supports of the outer sphere, the device for moving the test charge and the device for measuring the force on it are not too bulky, the field will then not be greatly distorted.

Having spent many laborious weeks constructing the apparatus, the experimenter takes the unusual step of retiring to his study to think.

It occurs to him that every tube of unit flux that originates on the inner sphere ends on the outer one, and that every one of these tubes would pass through the surface of any intermediate sphere that was concentric with the two conducting ones. The area of each such sphere would be proporttional to the square of its radius, r, and so the flux density at a distance, r, from the centre would be inversely proportional to r^2. The force on the test charge that he proposes to insert will, he knows from his experiment with the parallel plates, as well as from the definition of 'flux', be proportional to the flux density, and so this force, too, will be inversely proportional to the square of the distance r, from the centre.

Having at long last reached this conclusion, our experimenter decides to abandon the experiment. He now knows what the result will be. The inverse square law will be confirmed all right. Five minutes of thought in his study have taught him more than he could learn from five weeks of trial and error in his laboratory.

Let us list the things that thought has taught him:
(1) A correct statement of the inverse square law must strictly limit its application. If it does not do so, it is misleading. It is necessary to begin the formulation of the law with the words 'In a uniformly divergent field, and only in such a field . . . .'
(2) It is incorrect to say that the law defines the nature of electric charge, magnet poles, mass, or any other source of a field of force. It defines the nature of one thing only, a sphere. Hence it is also incorrect to say that there are really three laws. The proper place of the law is not in electricity, magnetism, astronomy, mechanics or in any other such science. It is in geometry.
(3) It is incorrect to say that the law can only be known by experiment and observation. It can also be known by taking thought.
(4) It is incorrect to say that the law is just one of those scientific facts that cannot, and need not, be explained. It can be explained with the help of simple geometry.
(5) It is incorrect to say that one could not replace the three inverse square laws by one single generalization in which the source of a force was not mentioned, and that would, nevertheless, enable the student to infer how the force varies with distance. The generalization that the area of a sphere is proportional to the square of its radius would suffice. The student could do the rest by the use of deductive reasoning. This reasoning would be based on the definition of force and the fact implicit in the definition that action and action are equal and opposite.
(6) To say that the inverse square law is only a first approximation to the truth is the same as to say that the area of a sphere only to a first approximation is proportional to the square of its radius.

Therewith one may say good-bye to all hope of explaining the recession of the nebulae or any other large-scale cosmic events by inventing the *ad hoc* hypothesis that the inverse square law needs the addition of a further term. But I do not think that one ought to dismiss this law from one's mind when seeking to explain the cohesion of the atomic nucleus. Let us examine statement (5) above. Is it necessarily true that the area of a sphere is proportional to the square of its radius?

Undoubtedly, one is inclined to answer hurriedly: 'It follows from Euclid.' This is true enough; but do the spatial relations within the nucleus necessarily follow from Euclid? Is space within the nucleus Euclidean?

What I am suggesting is this. Instead of saying that charges within the nucleus do not obey the inverse square law, it would be both more precise and more

253

fruitful to say that space within the nucleus is non-Euclidean. Such a mode of expression would lead to exploration of the various ways in which the space contained by the nucleus could be curved, and might give insight into the nature of the nucleus and of the forces that hold it together. I suggest this theme as likely to prove rewarding to a research worker. If the curvature is very intense forces must, for geometrical reasons alone, be very far from proportional to the inverse square of the distance between the bodies in question.

It may, for all I know, be impossible to postulate curvature of such a kind that a force of repulsion in flat space would operate as a force of attraction in curved space. But this is not the only way in which the puzzle of nuclear coherence might be solved. I shall venture to suggest a different approach in Appendix H. It is one that seems to follow from the new theory of gravitation.

Appendix G

What is an Elementary Component?

Science is hydra-headed. The answer to one question raises others. If Symmetrical Impermanence solves some problems, it gives rise to many more. Two of them are: What is an elementary component of the material universe? How does its extinction affect the nucleus of an atom?

I can only suggest tentative answers to these questions and should like to draw attention to their difficulty and their significance.

I have taken care throughout this book to avoid the impression that an elementary component of the material universe is necessarily any of the elementary particles with which nuclear physics has made us familiar. It is partly for this reason that I chose this non-committal, if cumbersome, name. The constituents of matter that, according to Symmetrical Impermanence, originate and become extinct may, for all that can so far be proved to the contrary, be found among the protons, electrons, neutrons, and other charged or uncharged particles that have so far been studied. They may, of course, be smaller. Nothing has so far been said to refute the hypothesis, if anyone cares to form it, that they are sub-particles and that the familiar particles are structures produced by a process of synthesis. But it is implicit in Symmetrical Impermanence that the elementary components are basic.

A few other considerations deserve brief mention.

Origins and extinctions should be expected to leave the average ratio of mass to charge for the whole universe unaltered, for if they did not this ratio would change with time. This suggests that whatever originates and becomes extinct as a unit is neither mass nor exclusively charge, but either a combination of both or something common to both.

Some objections to identifying those constituents of matter that I have called elementary components with protons, neutrons or electrons only have been mentioned in Chapter 3, Section 4. Mr. Wilson, of the Research Laboratories of the Central Electricity Generating Board, and Dr Sciama, of Cambridge, have drawn my attention to another objection to doing this. It leads to the second question, how does the extinction of an elementary component affect the nucleus of an atom?

If an elementary component were a neutron or any charged particle and became extinct in the complex nucleus of a radio-active substance, the balance of forces in the nucleus would be upset. The consequence would be the emission of other particles and of radiation, while what was left of the nucleus would become that of a different chemical element. There would, in short, be characteristic radio-activity. There would also be a residue of the chemical substance into which the disintegrating atoms were converted.

If the extinction of elementary components were the only cause of such radio-activity, the half-life of the substance would equal the half-life of matter. If there were additional causes, the half-life of the substance would be shorter. But it could never be longer than the half-life of matter.

Now good reasons have been found in preceding appendices for believing that the half-life of matter is likely to be not more than 4×10^8 years. If what became extinct were a proton or a neutron, no substance could have a radio-active half-life greater than this value. But many substances do have substantially greater radio-active half-lives. The radioactive half-life of most substances is indeed to all intents and purposes infinite.

From this it is clear that not every extinction of an elementary component causes radio-activity. It may be that none does so. But if some do cause radio-activity, it can be only a small proportion. It would not be so if the elementary component were a neutron or a proton forming part of a nucleus.

As mentioned above, the elementary component must have something that is common to mass and charge and so its extinction must result in the immediate, or delayed, disappearance of both from the nucleus. Therewith the loss of stability would not be as great as it would if mass or charge only were to disappear. It is worth investigating whether every nucleus that suffered such combined loss would necessarily show radio-activity. If the balance of forces was not greatly disturbed and the probability that further particles would be emitted were small, there would be no reason why the radio-active half-life should not greatly exceed the half-life of matter.

But the further difficulty mentioned above would remain. What was left of the nucleus in which an extinction had occurred would be the nucleus of a different chemical element. Extinctions must occur in nonradio-active elements and, according to this theory, a consequence of the continuous extinction of matter is a slow change of their chemical constitution. With a half-life of 4×10^8 years such a change would presumably be easily detected. If it does not occur, a different answer to the question has to be found.

One answer would be that a complete nucleus becomes extinct. If so, there is no chemical residue, no radiation, no emission of particles. An observable effect would be the pulse of gravitation that was emitted by the extinction. But the objection to this answer is that we regard the nucleus as a very composite affair, as anything but an elementary component. But I should not reject this answer out of hand. Conceptual distinctions do not necessarily coincide with real ones and it is not impossible that what is composite from the point of view of nuclear physics may be simple from the point of view of Symmetrical Impermanence. It has to be remembered that the individual constituents of the nucleus are not observed while they are forming a part of it, but only while they are outside. In Appendix H it will be shown that the nucleus of any element may well be a more basic unit than is usually supposed.

If this is accepted many puzzles concerning the nucleus disappear. At the same time the question is answered satisfactorily why extinctions are not accompanied by radio-activity and why they do not leave a chemical residue. The answer is that the unit to become extinct is always an entire nucleus.

Appendix H

Space and Mass

H.1: *The Physical Properties of Space*

When, early in this century, relativity theory made its tremendous impact on scientific thinking, one of the most revolutionary and profound conclusions to which it led was that the relation between space and matter had to be re-defined. As I have emphasized in Chapter 21 and elsewhere, we had until then thought of the relation as that between container and content. The container, space, was assumed to have no property other than extension and all distinguishing features that differentiate one thing from another were thought of as belonging to the content.

The content was thought of as comprising various constituents. There were, of course, ponderable matter and radiation. But not only these. Any given quantity of ponderable matter has an environment, and in this distinguishing features are provided by fields of force. These are yet another component of the material universe and were regarded in pre-relativity days as a part of the content that is accommodated in the container, space. They must occupy our attention for a short while.

Fields of force are measured and described in terms of their intensity, which is more precisely called their potential gradient. There are three kinds of field, electrostatic, magnetic and gravitational; so the environment of a given quantity of ponderable matter may contain three kinds of potential gradient. Each of these may vary both in magnitude and direction.

The potential gradients have a physical effect on ponderable matter, namely that of accelerating it; so they are physical realities. It follows that the environment of a given quantity of ponderable matter has physical properties.

In pre-relativity days these properties could not be attributed to space, as this was considered to have no property other than extension. So they were attributed to something else in the environment, something called luminiferous ether. This was regarded as a kind of matter and as yet another of the things that were accommodated in the container called space.

Like space, the hypothetical ether was endowed with the property of extension but was believed also to possess further properties, among them elasticity. By

258

virtue of these it was supposed to be subjected to a strain in places where there was a potential gradient, the amount of strain being proportional to the gradient.

This view has had to be abandoned. One of the pieces of insight gained from relativity theory is that it means nothing to speak of a con- tainer for the whole material universe. The notion that space serves this one and only function is, therefore, not held today. We know now that such space would be but a meaningless abstraction.

Nevertheless, a name is needed for the physical reality that surrounds any given quantity of ponderable matter. 'Luminiferous ether' might have been retained for this were it not that its verbal currency had been depreciated by sundry misapprehensions about it. Hypothesis had, for instance, endowed the ether with a variety of properties, of which elasticity was but one; density, mass and others had also been spoken of. But no such properties could be attributed to places where there was no ponderable matter. The only observed, or even inferred, physical features of the environment were various potential gradients, and so the many properties that had been attributed to the ether proved both inappropriate and too numerous to make this term suitable. If the word 'space' suggested some- thing too abstract, the words 'luminiferous ether' suggested something too concrete.

Perhaps the non-committal word 'environment' might have been a good choice. It could have been justified scientifically; for it is strictly accurate to say that a quantity of ponderable matter is surrounded by an *environment* that has physical properties and one has to be very cautious about saying anything more than this. But to use 'environment' in this way would have been to raise the word to the status of a technical term and to use it, moreover, in an unfamiliar context. It would probably never have succeeded in ousting the established word 'space'. So I think that Einstein showed a sound intuition when he retained this word while changing its meaning from that of a featureless container to that of a synonym for featured environment. But when the word 'space' is used in relativity theory with its new and precise meaning, it does not at the same time represent the vague concept that it does in everyday speech. It has become a technical term.

Another piece of insight gained from relativity theory was that the properties previously attributed to the luminiferous ether could not be attributed to the environment of ponderable matter. They were all replaced by one single property described as 'curvature'. It was from the identity of inert and gravitational mass that Einstein arrived at the conclusion that a field of force, at least when the force is gravitational, is a region for which the geometry of space is non-Euclidean, and he used the expression 'curvature' as a means of describing the departure from Euclidean geometry of the space in which there

259

is a potential gradient. This, too, became a technical term with a unique and precise meaning in relativity theory.

The conclusion startled the scientific world. Until then we had regarded curvature as a purely geometrical property. That it could be regarded as a physical one caused scientists to revise their notions about the nature of matter. How, it came to be asked, can something as apparently abstract as a curvature of space interact with something apparently as concrete as a material particle? The question is as puzzling today as ever it was.

The best answer will eventually come, I am inclined to think, from a revision of our notions about what is abstract and what is concrete. In making a distinction between the things to which we think the two terms appropriate we are enslaved by the organs of sense perception with which we, as human beings, happen to have been endowed. We call those things concrete that we can perceive directly with the help of these organs and those things abstract that cannot be so perceived. But this is a surmise and I do not propose to pursue this line of thought any further here.

Relativists have succeeded in showing good reason why gravitational potential gradients should be identified with a condition of space appropriately called its curvature. But they have not yet succeeded in showing the same for electrostatic and magnetic potential gradients, though they still hope to do so. The effort to do this is called the search for a unified field theory, and I have briefly alluded to it in Chapter 21. Anything that I, or anyone else, can say today about the relation between space and matter may have to be modified if and when the search has succeeded. For this reason, if for no other, what is said in this appendix must inevitably not only leave some insistent questions unanswered, but also be very tentative.

Should it ever be proved that electrostatic and magnetic potential gradients are regions of curved space, we must expect the kind of curvature to differ basically from the kind that is identified with a gravitational potential gradient, for there is no interaction between gravitational fields and the other two kinds. A change of the charge on a body having inert mass does not change the behaviour of the body in a gravitational field. A change in the potential gradient of a gravitational field does not affect the forces between electric charges placed in it. Gravitational masses that do not carry electric charges and are not magnetically polarized fall with the same acceleration through an electrostatic field (provided there is a gravitational one to fall in) irrespective of its intensity. In other words, electric charges behave in the same way in gravitationally curved as in flat space.

From such observations it has to be concluded that, if electro-magnetic potential gradients are, like gravitational ones, curved regions of space, the geometry by which the curvature can be defined can hardly be of the same

260

Riemannian kind as represents gravitational potential gradients. This independence from each other of the different kinds of field takes some accounting for and even leaves room for doubt whether the electro-magnetic field is, as seems so reasonable to suppose, a region where the geometry is non-Euclidean. But I have nothing to contribute to the search for an explanation and do not propose here to discuss the relation between space and charge or that between space and magnetism, but only the relation between space and mass. I have no choice but to discuss this relation as though the only condition of space by which one region is distinguished from another is curvature and the only known kind of curvature is that identified with the gravitational field. The incompleteness of such treatment is made obvious by the known facts about electricity and magnetism. But all that one can do in the present state of ignorance is to see along what path the incomplete treatment takes one and to hope that better knowledge of the relation between space and charge will not necessitate too great a change of direction from that in which the chosen path leads.

H.2: *Interaction between Space and Mass*
I want here to define the limits to our present knowledge of the relation between space and mass. Relativity theory has contributed a great deal to that knowledge and I am hoping that Symmetrical Impermanence can contribute a little more. But there is still much about which we are ignorant and questions to which the search for answers is likely to prove rewarding come under two headings: the action of space on mass and the action of mass on space.

The Action of Space on Mass: Relativity theory tells us something about the motion of a particle to which no force is applied. The motion depends, according to the theory, only on the curvature of the space in which the particle finds itself. If the space is flat, the particle moves with a constant velocity, which may of course be zero velocity. If the space is curved, the particle moves with a non-uniform velocity; it experiences an acceleration or a deceleration.

If a force is applied to the particle, its movement is no longer wholly determined by the geometry of space. A billiard cue can accelerate a billiard ball in flat space and a shelf can, by exercising a force on the stone resting on it, prevent the stone from following the curvature of the space-time in which it finds itself.

I mention these well-known facts only to bring out what is and what is not known about the action of space on mass. Relativity theory defines the property of *space* by virtue of which it is able to act on mass; this property is technically called curvature. But relativity theory does not tell us anything about the property of *mass* by virtue of which an unrestrained particle follows the curvature. The property is given the name inertia. But to name a thing is not to explain it or to give any sort of information about it. Relativity has some important things to say about space but none about mass. Let me illustrate our present ignorance about the relation between these two concepts with the help of an analogy.

When one sees a trarncar turn a corner, one may give the perfectly true explanation that there are rails for it to travel on. But it is an insufficient explanation. It says something about the street, but nothing about the tram. To make the explanation complete one has to add that the tramcar is provided with flanged wheels, so designed that they fit the rails.

Similarly, when one sees a stone fall, one may give the explanation that the space in which the stone happens to be is curved and thereby causes a curved track in space-time. One has then found the equivalent of the rails that take the tram round a corner. The analogy is admittedly not perfect, for a force is exerted between the tram rails and the flanges on the wheels, whereas the stone follows the curvature of space without any force being exerted on it at all. But the analogy serves, nevertheless, to show in what way the relativistic explanation remains incomplete. It fails to include the equivalent of the trarn wheels. What, one is led to ask on receiving the relativistic explanation, is the feature of a particle that causes it to 'engage' with space, as it were, so that its track in space-time follows the curves? In the example of the trarncar, steel wheels run on steel rails. When an unrestrained particle moves in space, something is said by relativists to run on curvature. What is it?

I shall return to the question later. For the moment I want to formulate some others.

The Action of Mass on Space: For the sake of convenience I shall again repeat a few well-known facts. According to Newtonian mechanics a massive body causes other bodies in its vicinity to be accelerated. To the question why this happens the answer is that a field of force surrounds the body. But to the question: what sort of a thing is this field? Newtonian mechanics has no answer; it can at most provide a name. Newton did not attempt to explain gravitation; he was content to postulate it.

It is here that relativity theory steps in. It does say what sort of a thing a gravitational field is, namely a region of curved space. The answer is justified both because it is methodologically sound and because it has considerable explanatory power. But it leads to the further question: why should a massive body cause the space around it to be curved? Again, relativity says something about space, but nothing about mass. It is content to postulate the effect of mass on space without explaining it.

It is here that Symmetrical Impermanence, in turn, steps in. The answer that it gives has been developed in detail in Part Four. It is that the extinction of an elementary component of the material universe is accompanied by the

262

extinction of some space and that this latter gives rise to a wave of contracting space, manifest as the kind of curvature that acts on mass. This answer leads, in turn, to another question: why should the extinction of an elementary component be accompanied by the extinction of some space? Symmetrical Impermanence does not attempt to explain this coupling of space with mass; it is content to postulate it.

There is admittedly some justification for doing so. If matter were to become extinct while space did not do so, the average density of matter in the universe would decrease without limit; as it also would if space were to originate while matter did not. It was acceptance of this idea that space is originating (the expanding universe theory) coupled with the postulate that the mass density of the universe remains constant that led Hoyle, Bondi and Gold to postulate the hypothesis that became known as that of 'continuous creation'. Similarly, the average density in the universe would increase if space were to become extinct while matter did not, or if matter were to originate while space did not.

But if the assumption of this kind of coupling of space and mass can be *justified,* it has not yet been *explained.* Like relativity theory, Symmetrical Impermanence accepts a two-way interaction between space and mass without doing much to explain it. Yet there can be hardly any doubt that an explanation, if found, would add much to our understanding of the nature of matter.

H.3: *Some Questions about the Nature of a Particle*
The ignorance to which I have just been drawing attention is all about the nature of a particle. To illustrate how great this is I give below some sample questions. At the present stage of science satisfactory answers cannot be given to any of them. There is no more than a somewhat frustrated groping. Some of the questions have just been formulated above, some are well known, some have a particular bearing on the theme of this book. Each of these questions needs to be pondered over. None can be dismissed lightly.

(1) What in the nature of a particle causes it, in the absence of restraint, to follow a track that is determined by the curvature of space?
(2) What in its nature causes the particle to depart from this track when it is subjected to an impact?
(3) Why is the outline of an elementary particle indeterminate in such a way that the particle has some of the properties of a wave?
(4) What causes a particle to cohere with others in an atomic nucleus with a force sufficient to overcome the forces of repulsion between the positive charges in the nucleus?
(5) Why should a particle that collides with a nucleus sometimes disrupt the nucleus and sometimes be captured by it?
(6) Why does the ratio of mass to charge in a stable nucleus increase with increasing atomic number?

(7) How can, as is postulated by Symmetrical Impermanence, some of the mass of a substance become extinct without either producing radio-activity or leaving a residue of atoms of lower atomic number?

(8) Why are origins of space and of mass coupled and, similarly, extinctions of space and of mass coupled?

During the many years while I have been exploring the implications of Symmetrical Impermanence I have necessarily given much thought to all the questions in the above list. I could not, in particular, afford to ignore questions (1), (2), (7) and (8). If the facts that lead to these questions are not true, most of what appears in this book is invalid; so I should have liked to discuss only those four here and say nothing about the others in the list. But I have found that such isolation is not possible. The questions are so linked that the answer to one seems to lead to answers to the others. Hence all or none must be discussed.

Therewith I find myself on the horns of a dilemma. It has not been easy to decide whether I ought to publish here and now my tentative conclusions about the nature of a particle or not.

The main argument against doing so is a strong one. It is that I have not given the subject the same number of years of careful thought that have gone to the conclusions presented in Parts One to Four of this book. There I was able to introduce statements with such words as 'it follows that'. The statements were, so far as I could make them, inferences. If, after they have been scrutinized by others, they prove to be wrong, it can only be because of my faulty reasoning. But if I now discuss the nature of a particle I shall be obliged to introduce some statements with the much deprecated word 'perhaps'. If these statements are found to be wrong, it may not be only because of my faulty reasoning; it may be because of a less excusable indulgence in speculation. In other words, I cannot promise entirely to avoid *ad hoc* hypotheses.

Nor can I promise anything that deserves the title of a theory. The conclusions at which I have so far arrived take me no further than to the beginning of a path that seems to lead in a promising direction. It could be urged with considerable justification that I ought not to invite others to tread this path until I have explored it further myself.

Be it added that the nature of a particle is a very big subject. It is much too far-reaching to be adequately discussed in the final appendix to a book concerned with quite different problems. There are indeed good arguments for postponing discussion of this subject to a later occasion.

But there is the other horn of the dilemma. The relevance of questions (1), (2), (7) and (8) to the theme of this book has already been mentioned. Question (7) has been discussed in Appendix G and question (8) is an immediate offspring of the new theory of gravitation. Thus both these questions are new; and new questions have a peculiar insistence. While one is giving one's attention to the implications of Symmetrical Impermanence, these two questions tend to assume a significance greater than any of the others. It may well be an exaggerated significance. With a proper sense of proportion one would, I think, give pride of place to other questions in the list. But in the immediate context these two ought not to be ignored. Some may, I fear, regard it as a waste of time and effort to give any attention at all to the conclusions reached in this book unless someone can first provide them with answers to questions (7) and (8). These two questions may, in other words, provide a pretext for ignoring my main thesis, whereas they ought really to act as a challenge and stimulus to explore the subject further.

This consideration has eventually, and after several changes of mind on my part, outweighed my disinclination to publish conclusions that are still tentative. So the remainder of this appendix will be concerned with a highly condensed presentation of a vast subject. I shall try to show that there is quite a good prospect of taking the unification of physics a stage further by giving concentrated attention to the eight questions listed above.

H4: *Bound and Free Curvature*
The kind of electrostatic field that is measured by the force between charges is bound to the charge at its centre. An electron carries such a bound field around with it. So long as a conductor does not move in space and its charge does not change, the electric potential gradient at a given place does not change with time. If the electrostatic potential gradient is a region of curved space, it is one where the curvature can be called 'bound'. This curvature is represented to the imagination as a sort of halo surrounding the electron or proton and it will be convenient for the present purpose to use this term. I do not do so in a derisory sense.

An electrostatic field can also be free of a particular charge. It then travels, in partnership with a magnetic field, in the form of radiation. This happens when the two fields break off from the antenna of a wireless transmitter. As the radiation passes a given place, the electrostatic potential gradient there changes with time. One can call it 'free'.

The question now arises whether gravitational fields are analogous to electrostatic ones in that they occur in both the bound and the free condition. If so, one can speak of bound and free curvature.

According to the traditional relativistic theory of gravitation, it is not so. Only bound curvature is recognized. A neutron is supposed to carry a very faint halo of gravitational curvature around with it, analogous to the strong electrostatic field carried as a halo by the proton. This latter is supposed to have two halos, the faint gravitational one in addition to the strong

265

electrostatic one. The gravitational curvature is assumed never to occur as a free travelling wave, but always as bound to the mass at its source.

According to the new theory of gravitation that has been presented in Part Four this is denied. The neutron has no gravitational halo and the proton has only the electrostatic one. For gravitation does not come into existence until the object at its source becomes extinct. The field in which a stone falls, it will be remembered, consists of a succession of pulses of contracting space, which are the consequence of extinctions and travel outwards from the source of the field. Thus the gravitational field is regarded in the new theory as analogous to free electromagnetism, as analogous to the light that emanates from a lamp. Each photon that contributes to the illumination is detached from the lamp and travels freely.

Thus the traditional theory recognizes only bound gravitational curvature and it has been made to appear in Part Four that gravitational curvature is always free. But is this correct? If one accepts the new theory of gravitation, must one assume that the kind of gravitational field in which a stone falls is the only one? Or is it consistent with the new theory to assume that bound gravitational curvature is a physical reality and occurs in addition to free curvature? Are there gravitational fields that are not composed of waves and in which the potential gradient does not change with time?

I have already said that I shall permit some 'perhapses' to enter this appendix, even though I have hitherto deprecated them and have done my best to keep them out of the main parts of this book. Let the first 'perhaps' be one that occurred to many of us in the early days of relativity: perhaps an elementary particle is, apart from its charge, a region of gravitationally curved space and nothing else.

Eddington made this point a fundamental one. 'Mass is curvature', he said somewhere. At the time when he was writing there was a much used catch-phrase: 'Man is a kink in space-time'. Preoccupation with this hypothesis is less insistent today, but not because we are any nearer either to accepting or discarding it. It is only because we are today more concerned with the way particle acts on particle than with the way space acts on particle. The great attention paid to nuclear physics has diverted attention from the fundamentals of relativity. But I do not think that what is at most a slight lack of topicality makes the question whether mass is curvature or something else any less rewarding now than it has been in the past.

To avoid the complication of electricity and magnetism I shall begin by considering only a neutron. What is it made of? Must we postulate a substance that is different in nature from the technical concept, space? Shall we find ourselves obliged to speak of 'particle stuff'? Or shall we say, with Eddington, that the neutron consists of curvature and nothing else? When the question is

266

put with this disconcerting candour one is inclined to dislike every answer that can be suggested. But I think that one will have the greatest dislike for the suggestion that there is something deserving of such a title as 'particle stuff'. One will be inclined to accept the Eddingtonian answer, if only as the lesser evil. The theory that the neutron consists of curvature seems better to meet the Principle of Minimum Assumption and to offer a better prospect of further unification of physics.

If this is accepted, the volume occupied by a neutron is a region of bound gravitational curvature. Within this region the potential gradient does not change with time in the way it does when the curvature takes the form of a travelling wave of gravitation.

To say this is probably to say much the same as was almost taken for granted by relativists forty years ago. But there is a significant difference. Traditional relativity had to be a little vague about the nature of the curvature. It has to be remembered that it postulates what I have called a faint halo around the neutron, a region of gravitational curvature with unlimited extent and flattening with distance in accordance with the inverse square law. So far as I know, there was no discussion as to whether the mass of an elementary particle was this faint halo and nothing else, or whether the particle was a composite affair, consisting of the halo in combination with an inner core of more intense curvature. In those days the time was not ripe for discussion of this question.

To be consistent with the new theory of gravitation one has, however, to face it. As the new theory denies that a gravitational field surrounds every neutron, one must regard the neutron as being a simple, and not a composite structure, as consisting of intense curvature in the volume occupied by it and as free from the feeble curvature previously postulated in its environment.

Therewith we arrive at a new way of distinguishing between space and mass: Electromagnetism apart, space consists wholly of free curvature and mass consists wholly of bound curvature. For the word 'curvature' one can, without change of meaning, substitute potential gradient. A way of understanding this will be presented later in Section H.7.

The implications of this conclusion are manifold and I have no doubt whatever that they deserve prolonged study, even if such study eventually leads to rejection of the hypothesis that mass is curvature. But this is not the occasion for a detailed inquiry. I must be content to mention one implication very briefly here and some others in the following sections.

I have said in Section H.2: 'In the example of the trarncar, steel wheels run on steel rails. When an unrestrained particle moves in space, something is said by relativists to run on curvature. What is it?' If mass is curvature, the answer is found. Only curvature can run on curvature. It seems as reasonable a conclusion as one may hope for so long as one accepts general relativity.

H.5: *Space and Anti-Space*
If the curvature of which a neutron consists were unfolded, how much space would it occupy?

Space and mass are so coupled, it has been mentioned repeatedly here, that when some space becomes extinct so does some mass; and when some space originates some mass originates also. There is thus an equivalence between a given quantity of space and a given quantity of mass. It should be possible to arrive at the number by which these quantities are related.

The coupling of space and mass has been justified here, it will be remembered, by the argument that the density of matter in the universe would tend either to infinity or to zero if there were no such coupling. To maintain constant mass density the quantity of space that becomes extinct whenever a neutron does so must be the quantity that is, on the average, occupied by the mass of a neutron. Estimates of this quantity vary widely and are difficult to arrive at with any degree of certainty. A typical figure for the density of the universe seems to be the mass of one nucleon (proton or neutron) per cubic centimetre. Be it then said as a rough approximation that when the curvature constituting a neutron becomes extinct, the volume of one cubic centimetre does so too.

What is the connection between the two extinctions? Does the one trigger off the other, or is there a conversion from one thing into something else?

The triggering-off hypothesis would require for its elaboration the assumption of some rather complicated mechanisms; it would not meet the Principle of Minimum Assumption. So I prefer the conversion theory. But what is converted into what?

If one assumes that a quantity of curved space constituting the neutron is converted into a quantity of flat space, that there is an 'unfolding' of the space of which the neutron consists, then a cubic centimetre of space would not become extinct when a nucleon did; there would be an addition to the space that existed already. The conversion that we have to look for is one into something that cancels existing space, not into something that adds to it.

This consideration leads me to venture a further 'perhaps'. It is that there are two kinds of space, suitably named respectively 'space' and 'anti-space'. Such a distinction would complete the duality that seems to be observed for all the basic concepts in physics: positive and negative charge, N and S magnetism, the electron and the positron, the proton and the anti-proton, matter and anti-matter, positive and negative energy. Perhaps the duality of space and anti-space could be shown to comprise in itself all the others.

It is a bold assumption but it has a good deal of explanatory power. An inference from it is that an elementary particle is not a curvature of space, but a curvature of anti-space. A quantity of this, say one cubic centimetre, is curved or folded into the volume of a neutron or a proton. What I have called the extinction of an elementary component of the material universe is then the unfolding of this quantity of curved anti-space. The flattening out is equivalent to the extinction of an equal quantity of space and is manifest as a wave of contraction.

The hypothesis is not complete unless it is applied to origins as well as to extinctions. To be consistent one must regard the origin of space, which is manifest in the expansion of the universe, as the consequence of the unfolding of a corresponding quantity of curved space. What unfolds must be something that could appropriately be called an anti-particle and differs from a particle in consisting of curved space instead of consisting of curved anti-space. The unfolding of bound curvature of space leads to the effect that I have called anti-gravitation in Part Four. The relation'. are presented in the following table:

	CONSTITUENT	
	Curved Space	Curved Anti-Space
Effect when bound Effect when free	Anti-particle Anti-gravitation	Particle Gravitation

I should shrink before the novelty of this conclusion if there were not considerable justification for it. But it seems to have the virtue of much unifying and explanatory power. It brings gravitation into a rational relation to space and it will appear from Section H.8 that it also helps to establish a rational relation between particles and space. The justification for the above table is, in fact, largely to be found in Section H.8

Considered quantitatively the assertion that a particle unfolds into a cubic centimetre or so of space (or anti-space) is rather startling. We all know that a curved line can be measured in two ways: round the curve and across the straight line that joins the ends of the curve; and we know that the first measurement gives a larger value than the second. We can accept the statement that curved space is analogous to a curved line in that its volume can be measured in two ways; and we can believe that the one measurement gives a greater value than the other. Relativity theory has accustomed us to this notion. But for the gravitational fields around the earth or the sun the difference between the two volumes is small. What many may find difficult to accept, I fear, is that space can be so intensely curved back on itself that the difference between the two volumes is that between one cubic centimetre and the volume that we assign to a neutron. Yet this is what I am suggesting.

To those who find this disquieting I should point out that if we are committed to the notion that a particle is curvature and nothing else we are also committed to the notion that the curvature is very intense. We ought not to expect a gentle curvature of space to manifest those properties (hardness, capacity for hitting things, and so on) that we observe in a particle.

The question arises whether anti-particles can be observed and, if so, what their properties are. It is worth asking, but it is arguable that one should not expect to observe an anti-particle. I should prefer to regard particles and anti-particles as occurring normally as couplets. Their properties would exactly cancel each other and they would have no effect of any kind on their environment. To all intents and purposes they do not exist so long as they remain as couplets. So it is meaningless to ask if there are few or many, or where they are. I am inclined to think that it is meaningless even to ask whether they exist, so long as their existence would have no effect on anything.

Occasionally, at random, and without cause, one of the two components of the couplet becomes free, the one that consists of curved space.

Becoming free means here collapsing into a wave of expanding space, into a pulse of anti-gravitation. The other component of the couplet, the one that consists of curved anti-space is left behind and now becomes manifest as a particle. One might say that it originates at this moment; for nothing was previously observable. But it might be better to say that the particle is uncovered when its companion, the anti-particle collapses. Uncovering renders it effective and observable.

Though, as I have just said, it is meaningless to say that the intact and unobservable couplet exists, it is meaningful, after the anti-particle has collapsed and the particle been revealed, to say that the couplet *has* existed. The point is a metaphysical one, but worth appreciating.

Before proceeding further I should like to suggest another possible way of interpreting as conversions what seem like origins and extinctions of matter.

It is a commonplace in general relativity that the value of π is not the same in a gravitational field as it is in flat space, the reason being that the field is a region of non-Euclidean geometry. For the field around the earth the departure from the value of π as calculated for Euclidean space is very small. It is only just observable for the sun's gravitational field. But these are fields where the curvature is free and the curved anti-space has gone a long way towards flattening itself out. The geometry in these regions is nearly Euclidean.

If a cubic centimetre of Euclidean space is crowded into the tiny volume of a neutron, the departure from Euclidean geometry within the neutron is very considerable. If one expresses its volume as $V = (4/3) x r^3$, one must assign to x a value that differs from π by a very large factor indeed. It would, I venture to suggest, be rewarding to apply the mathematics of relativity to a space that was so extremely non-Euclidean as I am claiming for the inside of an elementary particle. Our understanding of the nature of a particle would be greatly enhanced thereby.

There is one suggestion in particular that I should like to make to anyone prepared to tackle this job. It looks on the surface as though x in the above expression for the volume of a sphere in highly curved space must necessarily be very many times greater than π. But is a geometry logically possible and consistent with the facts for which x would be very many times smaller than π? If there is it would presumably represent a curvature that differed from the assumed one by a plus or minus sign, This would lead to a distinction between curvature and anti-curvature, which might usefully replace the distinction that I have suggested above between space and anti-space. Instead of saying that a neutron was a curvature of anti-space, it might be methodologically preferable to say that it was an anti-curvature of space. The hypothetical couplet that I have postulated as the unobservable parent of a simultaneous wave of anti-gravitation and a new particle would then be a couplet consisting of a combination of curvature and anti-curvature. As these would cancel each other's effects, the couplet would be literally indistinguishable from flat space.

One would then have even more reason to deny the existence of the couplet before its separation into two manifest components. An origin would be really the origin from nothing of a wave of free curved space which would leave behind it, as its counterpart, the region of bound anti- curvature that we call a particle.

In good time this would collapse and disperse as a wave of free anti-curvature, called a pulse of gravitation, and it would be as though nothing had been. After the passage of the two events only flat space would be left.

If this interpretation is tenable, there are, apart from electromagnetism, two kinds of basic process in the physical world. The first of these is the separation out of flat space of two things: one is a wave of free curvature and the other a minute region of bound anti-curvature. These are distinguishable only in that the one is free and the other bound. The second process is the subsequent collapse of the bound anti-curvature. In collapsing it appears as a wave of free anti-curvature, which is equal and opposite to the previous wave of curvature. It differs from it, however, in time and space. If the two waves coincided in time and space, there would be no effect. Physical events occur only, according to this metaphysical interpretation of the relation between space and mass, because two processes that are equal and opposite are separated by a space-time interval. The very existence of a physical universe depends on this random, indeterminate, and uncaused interval of time between the origin of a wave of curvature and the later cancelling wave of anti-curvature.

These are very bold speculations and I should be surprised if they will survive without drastic amendment. Such ideas ought not as a rule to be presented to others to work through before their author has done much work on them. But there are exceptional occasions when one is justified in suggesting a line of investigation to others rather than in keeping it to oneself, and I regard the present occasion as one of these.

For one thing it is important to show that question (8) is one to which it is worthwhile to seek an answer and my way of showing this is to produce possible ones. I should rather make suggestions that may have to be discarded than leave people with the impression that the question is too difficult even to tackle.

For another thing, I have wanted to illustrate the rather important point that progress in basic physics cannot be expected without careful attention to metaphysics. The scorn that is sometimes cast on this discipline is, I feel sure, usually misplaced. If the hypotheses that I have put forward very tentatively about curvature and anti-curvature prove invalid, I shall regard them as having served a purpose if they direct attention to the crying need for bold and, I must add with emphasis, metaphysical thinking.

H.6: *The Notion of a Graphical Symbol*
Much in the last section recalls what has been said in Chapter 23 about the distinction between deductive and representational understanding. A lot of space in a small volume may be understood deductively, but not representationally. The basic concepts of physics are, it will be remembered, by their very independence of any particular observer, beyond our powers of representational understanding. Any effort to understand them in this way is therefore misplaced.

This conclusion raises the question of models suitable for representing basic physical concepts. Lord Kelvin clung firmly to the view that such models are always possible. He even went so far as to deny physical reality to a concept of which he could not make a model. But in Lord Kelvin's day the need in physics to 'eliminate the observer', the importance of this kind of objectivity, was not yet appreciated. It has brought about a revolution in methodology. It could still seem quite obvious to contemporaries of Lord Kelvin that a statement about basic concepts in physics could only be true if one could represent it by the kind of model that would have a meaning for the human imagination. But it has now become clear that the opposite holds: the statement *cannot* be both true and basic if it can be so represented; for then the observer has not been wholly eliminated. What can be represented in the human imagination, it will be recalled from Chapter 23, depends on the particular organs of sense perception with which *homo sapiens* happens to be endowed. But a basic statement in physics must not depend for its validity on this fortuitous biological circumstance; it must be distinct from any of those things to which the human imagination has access.

This somewhat disquieting piece of methodology is relevant here because it applies forcibly to the relation between space and mass. Curved space cannot be represented by a model that the human imagination can grasp and neither can the process by which the curvature of space influences the motion of a particle in it. One may hope some day to understand the relation between space and mass deductively, but one can never hope to understand it representationally.

Yet it is only human to try to make models of all one's concepts and some have been made for elementary particles. They are by no means to be deprecated; for each has its use. Even if they suggest some things that cannot be true, they also suggest some things that are. When, for instance, one is concerned with collisions between elementary particles, one represents each particle by a model that is a small hard sphere. The imagination pictures this as round, elastic, slightly deformable (like a tennis ball when it bounces), and as unbreakably strong, although the word used for this latter property is the rather evasive one 'indivisible'.

Such a model is helpful up to a point and represents at least a part of the truth. But there are occasions when it has to be abandoned in favour of a very different one, a bundle of waves, perhaps. The various models contradict each other and it is right and proper that they should. But this means that the word 'model' is not well chosen. I should like to see it replaced by a different word. The limitations of efforts at representation would thereby be recognized.

May I recommend the term 'graphical symbol'. It is exactly what we often mean when we now say 'model'. We may represent an electron by a circle on the paper and by a sphere in the imagination, but we intend no more than a symbolic way of representing some features of the electron. When we think of a blurred outline for the electron we intend no more than a symbolic way of representing its wave properties. When we speak of curved space and represent it by a curve on a sheet of paper we intend no more than a symbol for the potential gradient that occurs in the environment of a given quantity of ponderable matter. When we speak of transverse electromagnetic waves we intend no more than a symbol for a periodic change with time that occurs at right-angles to the direction of propagation of the waves. Sundry statements in quantum mechanics come into the same category.

In such instances we are only too often misled by the symbol into thinking that it is a picture of reality, differing from the thing symbolized in nothing except scale. The term 'model' encourages us in our error. The term 'graphical symbol' would be a salutary corrective.

The recommendation that 'graphical symbol' should become a recognized technical term in the methodology of physics deserves a longer essay than there is space for here. My reason for introducing it is that I want in a moment to suggest a new way of regarding an elementary particle and I want at the same time to avoid the impression that this, or any other way, can convey the truth, the whole truth and nothing but the truth. I want to go further and insist that this way, like all others, sometimes misleads. I want to propose little more than a new graphical symbol, but one that will convey features of an elementary particle that are not conveyed by the symbols conventionally used.

H.7: *Point Symbols* A graphical symbol is needed to represent, not necessarily pictorially but symbolically, the notion that a particle is a region of curvature (or anti-curvature), that its boundary is approximately but not quite precisely defined, that it is accelerated in curved space and so on. The symbol shown in Fig.7A would serve. It needs a name and the shape suggests 'point symbol'. It is thereby distinguished from the more familiar circle, which could be called a 'sphere symbol'.

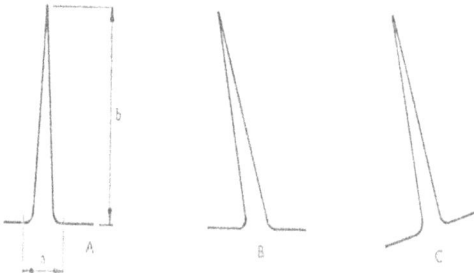

A. *In flat space and moving with uniform velocity; no force, no acceleration.*
B. *In flat space and accelerated by an applied force; force and acceleration.*
C. *In curved space and free from restraint; no force, acceleration.*

274

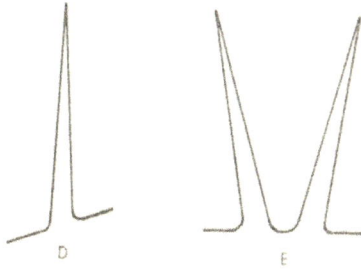

D. In curved space and prevented by a restraint from being accelerated; force and no acceleration.
E. Elastic collision between neutrons; mutual forces and accelerations.

Fig. 7. Point symbols of a neutron

In interpreting the symbol the following conventions apply:
Any horizontal line represents flat space and any departure from the horizontal represents a curvature of space. (In spite of the tentative suggestions in section H.5. I shall speak here of curvature and of space. That will not preclude the substitution of the word 'anti-curvature' or 'anti-space' later, if either is found to be preferable.)

The distance 'a' represents the volume of the particle. It is analogous to the distance in flat space between two points that adjoin the particle on opposite sides. As the point is shown to rise gradually from the horizontal, the diameter is not represented with precision; the graph therewith represents symbolically the known uncertainty about the diameter of an elementary particle. If the bend where the graph rises from the horizontal were analyzed in a Fourier series it would give a collection of sine functions. These are a graphical symbol for the waves that can, in certain contexts, represent a particle. As 'a' is a horizontal distance, it represents the amount of Euclidean space to an appropriate scale that the particle occupies and not the amount of space that it would occupy if the curvature were unfolded or smoothed out. This latter quantity is represented by the height of the point above the base line.

An analogy is a knot in a piece of string. This could, indeed, serve as an alternative graphical symbol; it could be called a 'knot symbol'. The distance across the knot along a straight hne between two points on opposite sides of the knot is quite short. But when one follows the twists and turns of the string between the same two points one measures a much greater distance. A knot can be described, a little paradoxically, as a long piece of string that occupies a short distance. The point symbol for a particle represents the notion

275

symbolically that the particle is a large amount of curved space in a small volume.

Let us imagine a person who studies a knot but can only measure distance along a straight line. He cannot get inside the knot and measure the length of string that is there. He can only measure the distance along a straight line that is taken up by the knot. In consequence his observations and measurement convince him there there is only a small length of string. It is only when the knot has been untied that he discovers the truth.

All analogies are imperfect, but this one serves to illustrate that it may mean something to say that a particle contains more volume than can be observed or measured. We approach the particle from the outside, from Euclidean space, and can ascertain thereby only how much Euclidean space there is between points on opposite sides of the particle. By this process we are unable to discover how much curved space there is between the same two points.

The value of a point symbol is enhanced if one adopts the convention that a tilt of the axis from the vertical represents an acceleration. This is shown in Fig. 7B. The shape of the graph is distorted by the tilt, which is a symbolic way of representing the fact that something has been done to the particle when it has been accelerated in flat space. Such a distortion can be thought of as occurring when the particle is hit by another one.

There is one particular feature about both a knot and a point symbol that might represent actuality but is more likely to be misleading. This is the suggestion conveyed by both symbols that space has dimensions additional to the three that we attribute to it. A piece of string has a two-dimensional cross section and is knotted into a third dimension. Hence a knot symbol suggests a space with one additional dimension. In the point symbol a horizontal distance represents volume in three spatial dimensions, each at right angles to the other two. By the same convention a vertical distance represents volume in three further dimensions, each at right angles to the other two and also at right angles to each of the three dimensions represented by a horizontal distance. The point symbol represents a six-dimensional space. Does this correspond to actuality?

It does not seem to. True, one cannot represent curvature in the imagination without picturing curvature into something. A line has one dimension and it curves into a second one; a surface has two dimensions and curves into a third. When one tries to picture transverse electromagnetic waves one is led to think of the electromagnetic field as curving into a fourth, and perhaps also a fifth dimension. The urge to interpret basic physical phenomena with the help of additional dimensions was not new at the beginning of this century. But Einstein made it more insistent when he introduced the notion of curved space. One conjured up extra dimensions for space to curve into in a desperate effort at representational understanding. But one should not forget that such efforts are misplaced.

Let me put what I am trying to say in a different way. Anyone who viewed the point diagram from below a line in the plane of the paper would see only the length 'a' and not the point. This length symbolizes a three-dimensional volume. But the whole diagram symbolizes a six- dimensional volume. To assume that actual space is six-dimensional would get us out of some difficulties. But I distrust easy ways out of difficulties and, besides, this one is not new and would have been taken years ago by those who are seeking a unified field theory if it were sound.

In this instance prepositions need to be carefully noted. Einstein did not speak of curvature of a three-dimensional body *in* space; he spoke of curvature of space. This is something of which the imagination cannot, and should not, hope to form a picture. Here again I make no apology for introducing some metaphysics. If an apology is owing it is for my inability to introduce more. More is needed in basic physics if understanding is not to lag so far behind knowledge that both will be lost. But I must return to the point symbol.

Rightly or wrongly this point symbol as shown in Fig. 7B represents the notion that there is an equilibrium state of curvature, for which there is some sort of symmetry. Disturbance of the symmetry is resisted and, on removal of the disturbing force, the previous condition is restored; Fig. 7a returns to Fig. 7A with cessation of acceleration. To suggest this is to attribute elasticity to an elementary particle, and there are reasons for thinking that one is justified in doing so.

The symbol would aptly represent the notion that a finite, though very minute, interval of time elapses between the moment when the distorting force has been removed and symmetry is restored. The graph is not a bad symbol for a particle that shivers like a jelly after it has been hit and in which the distortion can be accentuated if repeated impacts have a certain, resonant, frequency. Whether this is a wanted symbol or not, I should not like to say categorically. But resonance is a known phenomenon in nuclear physics and so this feature of the symbol may prove useful.

We are in the habit of thinking of inertia as the property by virtue of which a body in flat space resists a change in its velocity. Fig. 7B does not quite symbolize this notion. What it does symbolize is that inertia is an indirect effect. The immediate resistance of a particle that is subjected to a force is resistance to its distortion. The acceleration in turn is the consequence of this distortion, of the tilting of the axis. Here again, I think that the hint ought to be taken. It would be worthwhile to inquire whether the acceleration imparted to a particle by an impact is the direct effect that we picture or the indirect effect of a change in the way the curvature of the space constituting the particle has been altered by the impact.

Fig. 7C is a point symbol for a particle that is in a gravitational field and free from restraint. The tilt of the axis from the vertical again represents the acceleration of the particle. The lack of distortion represents freedom from restraint. The angle of the base of the graph to the horizontal represents the potential gradient in which the particle finds itself.

It should be noticed that the acceleration appears as such to an observer whose frame of reference is parallel to the edges of the paper, but an observer whose frame of reference was parallel to the base line of Fig. 7C would not think that the particle was being accelerated. He would be in the position of a person who is falling freely and observes a stone that is doing the same.

Fig. 7D represents a particle in a gravitational field when the particle is prevented from falling by a restraining force. The distortion symbolizes the restraint and the vertical position of the axis symbolizes the fact that there is no acceleration.

Fig. 7E is the graphical symbol for elastic collision between two particles of equal size. It shows the distortion that results for each from the collision. It is a measure of the force of impact between the particles. The opposing tilts represent the accelerations with which the particles rebound from each other after impact.

The point symbol, it has already been shown, represents elasticity, not as a property of a 'particle stuff', but as a property of curvature. By this convention the same symbol also represents the identity of mass and energy that was discovered by Einstein. When the curvature is increased the result is an increase of mass. But space resists a change in its curvature from a condition of symmetry. What is done to overcome this resistance is called a supply of energy.

What can be expressed somewhat vaguely and in qualitative terms only by graphical symbols can be expressed precisely and in quantitative terms by letter symbols, which are the basis of algebra. I am now suggesting that there is profit in taking the notion quite literally that a particle is a region of very highly curved space. Einstein took the notion quite literally that a field of force is such a region. He developed the notion with the help of letter symbols, and with most fruitful results. If the logic of algebra is applied with similar rigour to the curved space called a particle the result will be, I venture to suggest, equally fruitful.

Point Symbol for a Composite Nucleus

278

The picture that most of us form of a composite atomic nucleus is rather like a blackberry. We think of a number of minute, pip-like, spheres packed tightly together. Some of these represent neutrons, others protons. To distinguish them from each other we probably endow each kind in our imagination with a distinctive colour. If I were to give a name to this picture, I should call it a 'berry symbol'. I should certainly not call it a model. I doubt whether any nuclear physicist would claim that it was a true model of actuality.

A berry symbol has its uses, but it can also be sadly misleading. Let us consider some of its implications.

Each of the pips, be it a proton or a neutron and collectively called a nucleon, has mass. According to the traditional relativistic theory of gravitation it is surrounded by the faint halo of a gravitational field to which I have alluded already. The gravitational forces within the nucleus thus appear to an adherent of the traditional school as very faint indeed. The berry symbol represents symbolically a structure in which these gravitational forces provide a negligible amount of cohesion. According to the new theory of gravitation, be it noted by the way, the nucleon does not have even this faint halo of a gravitational field.

The same symbol shows the protons as very close together, as almost touching. But as these repel each other according to the inverse square law, the forces of repulsion between them are represented in this symbol as enormous. It is a symbol for a structure that contains little else but strong disruptive forces.

Accepting the implications of this symbol, some nuclear physicists have had resort to the *ad hoc* hypothesis of a new kind of force in nature, which is postulated as an addition to the gravitational and the electro-magnetic ones. The new kind of force is assumed in this rather makeshift hypothesis to operate with great intensity over very small distances only and to be one of attraction between nucleons. Its sole justification is that it is the best that has been found consistent with the traditional hypotheses about gravitation and the nature of a particle. Others seem to have adopted the *ad hoc* hypothesis that has been mentioned in Appendix F, namely that matter is permitted to disobey the inverse square law over very short distances.

I venture to suggest that neither the hypothetical nuclear force nor the hypothetical dispensation from the inverse square law would ever have been invented were it not for the misleading character of the berry symbol. I should like to show below that we have been mistaken in thinking that gravitational forces within the nucleus are very weak and electrostatic ones enormously stronger. There are good reasons for the conclusion that the gravitational forces greatly predominate in a nucleus of low atomic number and that the two kinds of force are of the same order of magnitude in a nucleus of high atomic number. Let me show why.

I have already drawn attention to the fact that electric charges behave in the same way in gravitationally curved space as they do in flat space. They are quite unaffected by the gravitational kind of curvature. Let me illustrate this by an analogy. A trumpet is a curved tube. But the curvature is introduced mainly for the convenience of playing. If the tube were straight, the trumpeter could not reach the stops. The pitch of the note that he blows is practically the same as it would be if the trumpet were not curved.

In this tube two nodes of a standing wave are separated by a certain distance. What is it? If the tube is bent back on itself, the nodes may be quite close to each other along a straight line. But a distance along this straight line is irrelevant. It does not affect the pitch. The curved distance measured along the tube is the one that counts for the determination of acoustic results. In this illustration the trumpet is shortened in such a way as to bring the stops within reach of the trumpeter's hands, and yet this shortening does not reduce the distance between nodes in a standing wave.

In curved space, when the curvature is of the gravitational kind, electric charges behave like standing waves in a curved trumpet. The curvature does not affect the separation between them. So the question arises whether the charges in a composite nucleus are all in such gravitationally curved space. If so, the nucleus could be represented by a knot symbol as shown in Fig. 8 for a helium nucleus. Each U-shaped element represents symbolically the quantity of curved space that is occupied by the mass of a nucleon, be it a neutron or a proton. The two nuclear charges that characterize helium are represented by plus signs. The volume that external measurement assigns to the nucleus is symbolized by the horizontal distance across the diagram. The volume of gravitationally curved space that constitutes the nucleus (perhaps six cubic centimetres) is symbolized by the labyrinth measured from end to end. To be a quantitative symbol the U-shaped elements would have to be very long and thin.

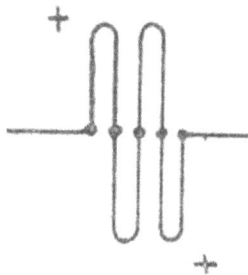

Fig 8 A possible graphical symbol for a helium nucleus

This knot symbol represents the notion that the nucleons are all fused into the same curved space. The berry symbol does not do so. It represents the notion that each nucleon preserves its identity. Which is the true notion?

So long as one thinks of an elementary particle as a hard, round, unbreakable sphere one will prefer the berry symbol. One will then have no choice but to invent nuclear forces or something else to explain the cohesion of the nucleus. But if one casts one's mind back to the early days of relativity and remembers the conclusion that a particle is curvature, one will be more inclined to accept the notion that when two curvatures come into very close proximity they fuse into one. The knot symbol will then appear the more appropriate.

In this there is no need to invent nuclear forces. The distance between positive charges has to be measured along the labyrinth, for charge ignores gravitational curvature. This distance separates the charges by a substantial amount. According to the inverse square law the force of repulsion between them is moderate. But all the nucleons are in a region where perhaps six cubic centimetres of Euclidean space are curved into the small volume of the nucleus. The curvature is very steep, fantastically so in comparison with the curvatures in the free gravitational fields with which we are more familiar. The cohesive forces of gravitation between the nucleons are thus enormous.

A point symbol illustrates the same features. One is shown in Fig. 9 for a boron nucleus.

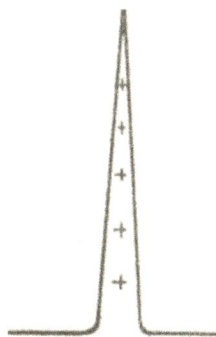

Fig 9 Point symbol for a boron nucleus

The horizontal distance along the base symbolizes the volume as generally stated, while the height of the point symbolizes the volume as it appears to the five charges. If the symbol were drawn to scale its height would be very great. It would then be clear that the charges were well separated from each other.

A few further implications of the point symbol are worth exploring. Every charge within the nucleus repels every other one. So each charge is subjected to a bigger force of repulsion if there are many other charges than if there are few. If charge density is defined as the number of charges per unit volume of the space that is curved or folded into the nucleus, it is easy to show that for constant charge density the force of repulsion on each charge increases with the number of charges present. Now constant charge density is obtained when the ratio of protons to nucleons is constant. So the tendency for the nucleus to disrupt under the influence of internal electrostatic forces would increase with increasing atomic number if the number of neutrons always bore a constant ratio to the number of protons. For the nucleus to have coherence one should expect this ratio to increase with the atomic number. Observation shows that it is so. The larger the atomic number the greater is the relative number of neutrons in the nucleus. This ensures that the charge density decreases with in- creasing atomic number and that the separation between adjacent charges increases also, giving thereby less force of repulsion between one charge and the others.

Let us return to Fig. 7E. It represents a collision between two particles of equal mass. If one of them is larger than the other and carries positive charges while the smaller one is a neutron, one has Fig. 10. The tilt of the neutron, and therefore its acceleration, is shown as much greater than that of the heavier particle; as it is in actuality.

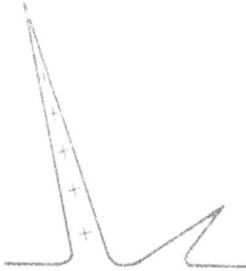

Fig 10 Collision between neutron and nucleus

The point symbol allows for other consequences of a collision between particles. This is, it must be remembered, not between things made of 'particle stuff' but between things made of 'curvature', if the expression be allowed. The first result of the collision is a distortion of each colliding particle. It is not difficult to believe that this distortion can be of various kinds, depending on the violence of the collision and, perhaps, on the length of time during which the colliding particles are close to each other. If the collision is comparatively gentle, it may cause elastic rebound. If it is more severe, it may lead to fusion of the colliding particles. This is called capture. When the nucleus is so large that the forces of repulsion almost predominate over the forces of cohesion, the collision may cause breakage, i.e., radio-activity and the ejection of particles.

This latter event can perhaps be regarded as a consequence of disturbing the relative positions of the charges in the nucleus. At rest, it is to be assumed, these take up positions in their many cubic centimetres of highly curved space where they are in equilibrium. But the distortion that immediately results from a collision displaces the charges so that some come closer together while others are more widely separated than before. Those that approach more closely then repel each other strongly enough for one or more to be ejected. Here the resonance that I have already hinted at (and that would be a consequence of the time lag between the production of a distortion and the regaining of the previous symmetry) may help to explain sundry phenomena in nuclear physics. But I have to resist the temptation of further exploring the ways in which a point symbol can suggest explanations for the observed action of particle on particle.

I do, however, feel under an obligation to say something about question (7). It is the question that was discussed more fully in Appendix G, namely, why an extinction can occur without both producing radio-active effects and leaving a residue of the nucleus of an atom with lower atomic number.

The question has answered itself in the course of this inquiry into more important aspects of nuclear physics. If it is true that a particle is curvature, that the components of the composite particle called a nucleus share a common curvature, and that the extinction of a particle is really the conversion of bound to free curvature, there is no difficulty. The nucleus is, regarded as a region of bound curvature, a single unit. If and when it collapses into free curvature, it does so as a whole. The extinction of some part of the nucleus can therefore never occur. It is all or nothing.

This conclusion raises the further question whether the probabilities of the extinction of positive and negative charge are necessarily exactly equal. If not an insulated body must acquire a small charge with time which might have observable consequences for a body of the size of the earth. There is, I suggest, a further field for research.

In this final appendix I have tried to give a fair sample of the difficulties that one encounters in basic physics. I have given prominence to those difficulties that are associated with various existing hypotheses about the nature of a particle and have mentioned difficulties that are inherent in my own theories as well as in those of others. I have tried to maintain a correct sense of proportion and to convey a true picture of the degree of tentative- ness or assurance that I feel about the various conclusions that I have presented.

Many more questions remain and as I write this it almost seems as though they were persons. Some tell me of fascinating avenues that they invite me to explore. Some remind me that I have already gone quite a little way towards providing them with some sort of an answer and urge me to proceed further. Some are baffling, some tiresomely clamorous, Some look at me a little spitefully and say that, if I do not satisfy them, they will prove that all my theories are invalid. But I must let them clamour and hustle. If I yield to these questions there will only be new ones. As I have said already, science is hydra-headed.

Epilogue

Towards a Solution

That some observations in physics can be explained as inferences from the Principle of Minimum Assumption does not prove that all can. The principle has been justified in this book by a few examples only and it is at least arguable that others would not have justified it. The principle may not be valid for the whole of physics; its unifying power may be incomplete; there may be laws of physics that are not implicit in it; these would be of the statute book kind. Some of the cosmic constants and some of the constituent parts of the material universe, again, may have to be attributed to a Cosmic Specification and not to mere randomness. Statute Book and Specification would then appear to be the works of a Creator in the respective capacities of Legislator and Architect of the material universe.

Many do take this view and regard any other as robbing our very existence of all meaning and purpose. At least some of the laws of physics, they feel impelled to believe, must have been devised with the purpose of creating order out of chaos. Some principle of design must have inspired the structure of the firmament. A building, they argue, has no beauty, not even coherence, if it has not been constructed to the specific requirements of its architect; unless one assumes that a Cosmic Specification has been imposed on the physical universe, one cannot explain the regularities that physicists tell us about; in a physical world that resulted from mere randomness nothing would be predictable.

Arguments can be urged both for and against this view, but nothing that has been said in this book can prove those wrong who hold it. It is, nevertheless, worthwhile to point out that those irreducibles with which the physicist of today is left are not of the kind that seems to support a teleological view of the physical world or to justify an idealization of matter. To the physicist they are immensely basic and important; from the point of view of religion, ethics and aesthetics they must look rather trivial. It would not be easy to argue that our existence would have either more or less purpose and meaning if the velocity of light had a different value. The question whether the charge on the electron, the quantum of action and the mass of the neutron are independent constants or implicit in each other can have no ethical significance, interesting though it is to science. The first law of thermodynamics has a very great value in the methodology of physics, but it does not belong to the same universe of discourse as those things that may be the fount of inspiration for an architect. What is today left in the Cosmic Statute Book and the Cosmic Specification has been listed in Appendix A. It is too meagre to provide a means of ensuring purpose, order, beauty, coherence, or even, I think, predictability. Those who hope to justify an idealistic view of the material universe have already lost so much that there is not much more for them to lose. They should be able to contemplate without any regret the prospect that the list of irreducibles may become ever shorter as science progresses.

Does this conclusion lead to a materialistic philosophy?

The question is not relevant here, but I have found it to be too insistent to be disregarded. It is not in human nature to reach conclusions about cosmological problems without considering their wider implications. So let me say that the method employed in this book, when applied consistently to the whole of reality, does not lead towards but away from materialism. As I have already written three books on this theme [1, 2, 3] very few words of explanation will suffice here.

A decision is called for between the materialist assertion that *all* events are the consequence of the unaided action of matter on matter and the opposed claim that *some* events are the consequence of the action, direct or indirect, of non-material influences on matter. It is this latter claim that I have defended in my previous books. In the present book I have gone no further than to discuss some of those events that do have to be attributed to the unaided action of matter on matter. They all belong to the physicist's universe of discourse and I should be prepared to place all other events that belong to the same universe of discourse in this category. This leaves the question open whether the physicists' universe of discourse is concerned with the whole of reality. It is my contention that it is not.

My reason for opposing materialism is simply that I have been led to the conclusion that the structure and behaviour of plants and animals, as also the works of man, have to be attributed to non-material influences. In other words I claim that reality has two aspects and I have discussed only one of these in the present book.

To judge between the two opposed schools one has to find the answer to a pivotal question concerning the nature of matter. Can the laws that govern the action of one material system on another - can, in other words, the laws of physics account for all the observations that we observe in the organic world and the works that result from man's activity?

There are other ways of asking the same question. Is the kind of order that we observe in the organic world imposed on matter by other matter or is it

286

imposed on matter by non-material influences? Is this kind of order the consequence of what matter does or of what is done to it?

Here I have only discussed structures in the rough untouched world of lifeless things. I have attributed these to the unaided action of matter on matter. But in doing so I have also shown that they can be accounted for without attributing to matter anything in the nature of a capacity for creating order, for selection, guidance, control, co-ordination. Therewith I have implied that it is erroneous to idealize matter as is done in the various materialist schools.

The true inference from what has been said here is that the kind oi order observable wherever things are touched by life must be attributed to the action of non-material influences.

1. *Science versus Materialism,* Methuen, 1940. (Republished 2010)
2. *Mind, Life and Body,* Constable, 1945. (Republished 2010)
3. *Facts and Faith,* Oxford University Press, 1955. (Republished 2010)

www.ingramcontent.com/pod-product-compliance
Lightning Source LLC
Chambersburg PA
CBHW020454100426
42813CB00031B/3358/J